ACKNOWLEDGMENTS

There are many people I would like to recognize for their contributions to this book. First and foremost, my editor, Judy Crowell Deasy, who also happens to be my beautiful wife, partner, and best friend. Without her love, support, and impeccable command of the English language, this book project would have been impossible.

Second, I always envisioned a book with a lot of photographs. I decided to use a melting pot of three talented photographers. Many thanks to Daniel DiGiacobbe, Krystyna Jutson, and John Parrish whose tireless efforts have certainly paid off in the beautiful photographs in this book. One picture is worth more than a thousand words!

Many thanks also to Tom Chatham. Chatham™, in my opinion, creates the finest created gems in the world, and it shows in the photographs that they allowed us to use in this book.

Lorenzo Yih and Danie Tangchai of LJ International provided those hard to find gemstones. Without their help, this book would be incomplete.

Daniel DiGiacobbe and Judy Crowell Deasy teamed up to do a fabulous job with the layout, and this project simply would not have been possible without them.

Last, and certainly not least, I would like to thank the thousands of television shopping viewers who over the years have constantly asked when I was going to write a gem book. It was certainly never my intention to write anything, but I am very glad that I did.

Paul Deasy

TABLE OF CONTENTS

INTRODUCTION

I have spent the better part of 17 years offering gemstones for sale on televised shopping networks. In all that time, it has occurred to me that the gemstone buying audience has an insatiable appetite for information on anything relating to gemstones. After countless viewer requests for book recommendations, I realized that the perfect book on colored gemstones had not been written. No, I am not suggesting that the literary world has been waiting with baited breath until I graciously decided to swoop in and save the day, but I am pointing out that most of the books on the subject have serious shortcomings in one area or another. First, most books on gemstones tend to be pretty technical in nature. I don't know about you, but after the second or third acronym, chemical composition, or Latin name I have a tendency to nod off. Second, most books give too much information. I suspect that if you are reading this book, you want a quick reference guide to help you learn something about your favorite gems, not a basis for research on your thesis in earth sciences. Third, the information in some books may not necessarily be up to date. Having visited various mines throughout the world, I was always amazed by what I learned that **wasn't** in any of the books I had read. Finally, most of the books are not organized in an easy to use format. Most are arranged by gem varieties, and you must look in the index in the back just to locate the page for the gem in question. As long as you know your ABC's, you will do just fine locating the gems in this book, as they will be listed in alphabetical order.

Here are some things you need to know before you start reading the book.

1. There will be a brief (I promise) section in the beginning on general information on gemstones.

2. There will be no information on diamonds except for a brief mention in the explanation of the Mohs scale of hardness and then again in the section on "cut." Diamonds are a completely different animal than colored gems and deserve their own book.

3. If a gem has a common name, that will be the one used for alphabetical order. In other words, tanzanite (the commonly accepted name for blue zoisite) will be found under "T" and not "Z."

4. While there will be information on care and cleaning listed for each gem, I can not stress enough the importance of having a suitable jewelry box for safely storing your jewelry.

5. Enjoy the book.

6. Recommend the book to all of your friends, and give as many copies as gifts as humanly possible. After all, I have two kids to put through college!

Whether you are just now becoming interested in gems, are already a gem collector, or think that your future may include a job in the fascinating world of colored gemstones, I sincerely hope you enjoy this book.

Paul Deasy

THE SCIENCE OF GEMOLOGY

The original study of gemology wasn't much of a science at all. As recently as the early 1900s, the basic premise was that if a stone was red, it was a ruby. If it was blue, it was a sapphire and so on. Of course, we now know that every gemstone has specific properties that make it unique. A red stone might be ruby, garnet, or spinel. A blue stone could be sapphire, tanzanite, iolite, or maybe even a diamond. So, how did we get to the science of gemology today? The National Association of Goldsmiths in London (now called the Gemmological Association of Great Britain) started a study in gemmology in 1908. By the way, that is not a misprint. That is how they spell gemmology. The first diploma in gemmology was handed out in 1913. An American by the name of Robert M. Shipley received his diploma in 1928, and shortly thereafter founded the Gemological Institute of America (GIA) in 1931. Today, the GIA is considered the world's foremost authority on gemstones.

WHAT IS A COLORED GEMSTONE?

A colored stone can be defined as any gem material other than a diamond. Pretty simple, isn't it? To leave it at that would be an injustice, so let's dig a little deeper. Get it? **Dig** a little deeper? .....Oh, never mind.

There are two basic types of colored gemstones: organic and inorganic. **Organic** means they come from living things—plants or animals. This is a short list and would include amber, coral, jet, pearls, shell, and ivory (it should be noted here that most ivory comes from endangered species, so trading in ivory from non-extinct animals has been banned in more than 100 countries). Most gems are **inorganic** and come from minerals. These are the non-living things like crystals and rocks.

THREE TRAITS OF ALL GEMSTONES

All gemstones share three traits: **Beauty, rarity, and durability**. Obviously, since all gemstones have specific physical, chemical, and optical properties that make them unique, they possess different levels of these three traits.

1. **Beauty**- This may be the hardest to define. After all, beauty is in the eye of the beholder. If you don't believe this is true, just look at the sweaters that some people wear! For the purposes of our discussion, let us agree that beauty in a gemstone comes from a combination of four factors-**color, symmetry, surface appearance, and transparency.**

    a. **Color**- This is the single most important characteristic in a colored gemstone. This sure makes it easy. If you really love the color of a gemstone, that's pretty much all you need to know.

    b. **Symmetry**- This is the balance or harmony of the cut of the gemstone.

    c. **Surface appearance**- What does the surface of the stone look like? Are there imperfections on the surface? Is it polished to a high luster?

    d. **Transparency**- This describes how much light can pass through the gemstone. Not all gems are equal in this area. Gems such as garnet or tanzanite are **transparent**. This means that light passes through with little or no distortion. Think of looking through a window pane. Gems such as opals or moonstone are **translucent**. This means light diffuses as it passes through. A frosted window or shower curtain would be an easy way to remember this. You can see light pass through, but you can't quite clearly see what is behind it. Still other gems, like onyx, are **opaque**. This means no light passes through. A brick wall is opaque.

2. **Rarity-** Some gems are rarer than others. When I ask which gemstones are the rarest, a lot of people immediately think of diamonds, rubies, emeralds, or sapphires. While these are indeed rare, they are not nearly as rare as stones such as alexandrite or black opal which can command much higher prices than many other stones. Rarity, in and of itself, does not necessarily mean it will command a higher price. Some gems are extremely rare but there is little market demand for them. Spinel is one good example. I think the value of this group would probably increase dramatically if it were ever aggressively promoted.

3. **Durability-** This may be the most misunderstood trait of them all. Most people think of hardness as the sole factor of durability. There are three factors affecting durability: **Hardness, toughness, and stability**.

   a. **Hardness-** This measures resistance to scratching. More specifically, it is the ability of one gem to scratch another gem. In 1822, a German mineralogist by the name of **Fredrich Mohs** developed a system for rating the relative hardness of minerals. Diamond is hardest at 10 and talc is softest at 1. This system is a bit confusing at times. Consider that a diamond (10 on the Mohs scale) is 140 times harder than a ruby or sapphire (9 on the Mohs scale)!

   b. **Toughness-** This measures how well a gemstone resists breaking and chipping. Topaz, while relatively hard (8 on the Mohs scale), breaks rather easily if you hit it in the right place. Conversely, jadeite, which is only 6.5 to 7 on the Mohs scale, is much tougher than topaz. The way it was first explained to me is an easy way to remember it. Peanut brittle is very hard, but if you drop it on the floor it is likely to shatter. It is hard but not tough. Caramel is very soft but will not shatter if you drop it. It is tough but not hard.

   c. **Stability-** This measures how sensitive a stone is to light, heat, or chemicals. For the purposes of this book, I will comment on this in the main section of the book only if a gem is particularly vulnerable to any of these elements.

If you learn nothing else from this discussion on durability, please commit this to memory. No gemstone is indestructible, and all gemstones should be treated with care and respect. Some people are under the mistaken impression that since a sapphire is 9 on the Mohs scale of hardness, and rated excellent on the toughness scale, that they will never have to worry about damage to the stone. While it is true that diamonds, sapphires, or rubies should wear better than a tanzanite or an emerald, that doesn't mean that they can't become damaged just the same.

THE FOUR C's

I suspect that anybody with a fleeting interest in gemstones has probably heard of something called the four C's. There are four criteria that are used to evaluate gemstones, and all begin with the letter C: <u>c</u>ut, <u>c</u>olor, <u>c</u>larity, and <u>c</u>arat weight. All four of these traits are vital, but when considering colored gemstones, color is the single most important factor that affects the quality and price of the gemstone.

COLOR

What causes color in a gemstone? Technically, it is something called "selective absorption." Although we see light as white, it is actually made up of waves of different colors. Any object, including a gemstone, will absorb certain waves and return others. The ones it returns determine what colors we see. But, I promised not to get technical, didn't I? As I mentioned before, all gems have specific properties that make them unique. However, the occurrence of certain trace elements will determine the color. Try this on for size. We get rubies and sapphires from the mineral group we call "corundum." Its chemical composition is aluminum oxide. Pure corundum is colorless. If the corundum is red, which we call ruby, the color comes from the appearance of the trace element chromium. If the corundum is blue, which we call sapphire, the color comes from traces of iron and titanium. This holds true for virtually every gemstone. What makes this even more interesting is that these trace elements can sometimes help an expert determine where in the world a gemstone might have been mined.

What color is the best?

One of the most common questions I'm asked is "what color amethyst (or ruby, sapphire, etc.) is the most valuable?" This is indeed a good question. Here is a pretty good rule of thumb. The deeper the color, the more valuable the stone. Remember, though, that beauty is in the eye of the beholder. Deep, royal colors of purple amethyst are the most expensive. If you happen to prefer lighter shades of amethyst, don't panic. You don't have to change your personal preference, it just means you should pay less money per carat for your amethyst.

CUT

The discussion of how a gemstone is cut may potentially be the most technical and confusing of all. Fortunately, we are not going to get into all of the parts, angles, and percentages of a standard round brilliant cut stone and other such sleep-inducing matter. There is a chart on page 194 that shows the most common shapes of faceted stones so you can recognize them. For the sake of simplicity, let's keep this discussion focused on what a colored gemstone cutter is trying to accomplish.

Diamonds versus Colored Gemstones

Diamonds and colored gemstones have completely different standards regarding how they are cut. With diamonds, we want to get the maximum amount of white light reflected back to the eye. Some dispersion would be nice as well. Therefore, a diamond cutter must pay strict attention to proportion, ratio, and finish. A colored gemstone is about finding the best color possible, so the precision of the cut is not all that crucial.

How does cut affect the other C's?

I should probably begin by saying that in the wholesale gemstone business, gems are sold by carat weight. People who buy **rough** (unfinished gemstones), don't ask how many stones, they ask how many carats and what is the price per carat. So **yield**, or how many finished stones can be obtained out of that rough, becomes very important. As the saying goes, "waste not, want not." Therefore, a colored gemstone cutter tries to maximize the carat weight whenever possible. That being said, how a gemstone is cut will affect all other value factors in a gemstone.

1. It may help alter the **color**. A cutter may cut a stone thinner to lighten a stone that would otherwise appear too dark. Conversely, he or she may cut a stone deeper to darken the color. If it is a **pleochroic** (pleochroism is when a gem shows different colors in different directions) stone like tanzanite or iolite, how they cut it will determine which color the eye sees.
2. It can affect **clarity**. A cutter can either eliminate inclusions or make them less noticeable by strategically positioning them.

3. It can affect **carat weight**. As we discussed earlier, a lapidary (stone cutter) may manipulate a gem, cutting it deeper so it appears darker in color. This may also affect the stone's weight making it heavier. The lapidary may also cut a stone wider than normal to give it the appearance of being a larger stone.

CLARITY

A major factor in any transparent gemstone's value is its clarity. Very few gems are totally free of inclusions. Incidentally, not many people in the trade say the word inclusion anymore. Just as our garbage man is now a "sanitation engineer" and a party planner is now an "event coordinator," the word inclusion is also going the way of the dinosaur. Now, **blemishes** and **inclusions** are called "**clarity characteristics.**" I won't go into great detail on the numerous types of clarity characteristics, but I would like you to know the two basic types. Blemishes are surface irregularities like scratches and nicks. These can become more prevalent with normal wear and tear on a gemstone. Inclusions are internal characteristics that can affect a gem in many ways. On people, a blemish would be on the skin, like acne or a scar. An inclusion would be inside, like a cracked rib or a clogged artery. Another reason for using these analogies is to further explain how these inclusions can affect the durability of a gemstone. A blemish on the surface of a stone will have little or no affect on durability. A fracture or break inside, however, can weaken a stone and affect its value. It is the same with people. That acne or scar may be unsightly, but it won't affect our durability. The cracked rib or clogged artery will definitely slow us down and make us less durable.

Eye Clean versus the Gem Microscope

If a tree falls in the forest and nobody is there to hear it, does it make a noise? If a gem has an inclusion and my eye can't see it, is it actually there? No, this is not my foray into Philosophy 101, but an attempt to talk about how gems are evaluated for clarity. When a gem is called **eye clean**, it means that no clarity characteristics are visible to the naked eye. In the jewelry business, gemstones are generally evaluated under ten power magnification. There are more powerful gemological microscopes available, but you generally don't want to use them. Why? If you detect imperfections under twenty or thirty power magnification that are not visible under ten power magnification, you would be obligated to disclose that information. If you are selling your home and it is in good condition and has passed all of the inspections, both you and the buyer will be happy. I doubt you would want to get some sort of sonogram to look through the walls to see if you can find any flaws in the two by fours behind the drywall. Likewise, if a gemstone is flawless under ten power magnification, why go looking for problems?

Clarity and Value

It is interesting to note that not all gemstones are evaluated equally with regard to clarity. Some gems, like aquamarine, amethyst, topaz, and citrine, tend to be eye clean. Emeralds, on the other hand, almost always show visible inclusions. This is taken into account when stones are evaluated for clarity. In other words, a visible inclusion in an emerald will not adversely affect its value very much. However, an aquamarine with the same level of inclusions would be much less marketable. Emeralds are like that high school science test that everybody in the class did so poorly on that the teacher had to grade it on a curve. In the jewelry business, emeralds are graded on a curve.

Believe it or not, inclusions can also have a positive impact on the value of a gemstone. For example, the star effect that can be found in some rubies and sapphires are actually needle-like inclusions that happen to be arranged in a particular pattern. Sometimes an inclusion may be a different gem material that gets trapped inside another gem as the crystal grows. Serious collectors might look for rare stones with particular inclusions. It is similar to a misprinted stamp for a stamp collector or a mistake that is made while minting a coin. Those errors can make the item even more valuable.

CARAT WEIGHT

Carat weight is exactly as the name implies. It is the measurement of the weight of a gemstone. A carat is a standard unit of weight for measuring stones, and is equivalent to 200 milligrams. You may also hear people in the jewelry business talking about **points**. A point is one one-hundredth (1/100) of a carat. If somebody says you have a three quarter (3/4) carat stone, it is the same as saying your stone is seventy-five points (.75). Where did we get the name carat? Many years ago, scales were nothing more than a simple lever that would balance one object on one side and another object on the other side. Carob seeds were used as the standard for weighing gemstones because they were pretty consistent in their size. One hundred carob seeds became the equivalent of one carat.

Carat versus Karat

Carat is a measurement of the weight of a gemstone. **Karat** is a measure of the purity of gold not the weight of gold. Unfortunately, some people mistakenly use the words carat and karat interchangeably. To make matters worse, some countries spell karat beginning with a "C," so the two words with different meanings are spelled exactly the same.

It is not my intention to cover precious metals in this book. However, since I had to introduce the term karat, here is a brief synopsis of the most common purities of gold. Gold in its purest form is twenty four karat (24kt) gold. 24kt gold is too soft to be used in jewelry and would be damaged much too easily. Other metals (copper, silver, and zinc) are added to the gold to give it strength. When two or more metals are mixed together it is called an **alloy**. Eighteen karat (18kt) gold is an alloy made of 75% pure gold and 25% other metals. Fourteen karat (14kt) gold is an alloy consisting of 58.5% pure gold and 41.5% other metals. Ten karat gold is an alloy that has 41.7% pure gold and 58.3% other metals. It is interesting to note that changing the amounts of those alloy metals is how different colors of gold are created.

Weight versus Size

Two different gems may be advertised as having the same weight but appear to be very different in size. Have you ever watched a televised shopping channel during a jewelry show? You might have seen them offering a ring style in your choice of five different gemstones, and while all the gemstones were the same physical size, the carat weights in each ring were different. Why is this? As I stated earlier, every gemstone has unique properties. For each gem listed in this book, you will see a number that indicates its **specific gravity**. This is also known as **density**. Specific gravity is the ratio of an object's weight to the weight of an equal volume of water. Sapphire has a specific gravity of 4.00, and water has a specific gravity of 1.00. Therefore, a cup of sapphires (assuming you could remove all of the air between stones) would weigh four times as much as a cup of water. The larger the specific gravity number, the more dense the stone will be. If you are comparing two different gemstones with the same carat weight, that means the stone with the smaller specific gravity number will be **bigger** than the stone with the larger specific gravity number. By the same token, if you put a five carat amethyst and a five carat blue topaz side by side, the amethyst will be larger in size because it has a **smaller** specific gravity.

Let's try to make this easier to understand. What weighs more, a pound of feathers or a pound of bricks? Obviously, this is a trick question, because they weigh exactly the same—one pound. However, the pound of feathers would take up much more **space** than the pound of bricks. The bricks have a higher specific gravity. Why is this specific gravity number important to a gemologist? Not only does the specific gravity help a gemologist identify gem material, but it has other practical applications as well. Suppose you damaged your one carat tanzanite stone in a ring and you wanted to replace it with a one carat iolite stone. Iolite is a great substitute for tanzanite because the look is nearly identical, but iolite costs far less. Looking at their specific gravities, tanzanite has a specific gravity of 3.35 and iolite has a specific gravity range of 2.57 to 2.66. The one carat iolite would be too big to fit in the setting created for the one carat tanzanite. This is why most jewelers order gemstones by dimension, rather than by carat weight, so they will fit into prefabricated settings.

Here are some more interesting facts about the practical applications of specific gravity. Amethyst is a very common stone in Brazil. It would not be uncommon to find some amethyst lying on the ground. That is exactly what a hunter in Brazil thought he had found when picking up a rock to put in his slingshot. The hunter thought that the purple stone was a bit heavy to be amethyst, so he brought it to the next town to have it tested. As it turned out, the specific gravity was between 3.74 and 3.94 (the specific gravity for amethyst is only 2.66). That hunter knew that the stone he picked up was too heavy for its size to be amethyst, and voila, Brazilian plum garnet was discovered. Here is one more example. The specific gravity for water is 1.00. If an item has a specific gravity of less than 1.00, it will float on water. Amber has a specific gravity range of between 1.05 and 1.09. Since the specific gravity of amber is so close to that of water, it will float in a solution of salt water. Much of the amber that comes from the Baltic Sea washes up on shore or is mined by taking a boat out and disturbing the floor of the sea so the amber floats to the top. If the specific gravity of amber were 4.00 like a sapphire or ruby, it would stay where it is, or at the very least be much more difficult to recover from the sea.

Weight and Value

In the wholesale side of the gemstone business, loose gems are bought on a **price per carat** basis. Using this price per carat basis is also helpful when trying to compare values of different gemstones. With certain gemstones, like rubies, sapphires, alexandrite, and tsavorite garnet, it is very rare to find good quality stones in large sizes. If you do find them, the price per carat increases dramatically. In other words, you would pay much more for a two carat ruby than you would pay for two one carat rubies of the same quality. Other gemstones like amethyst are much easier to find in good quality weighing ten carats or more. Therefore, you may pay the same price per carat for a five carat amethyst stone as you would for five one carat amethyst stones of the same quality. How can we use this information to help us? If you really like sapphires and you also prefer very large stones, then you need to either hit the lottery or consider blue stones that are much more readily available in larger sizes.

CARE AND CLEANING

This is perhaps one of the most important yet ignored responsibilities of jewelry ownership. Failure to properly clean your jewelry can have an enormous impact on the appearance and value of your gemstones. Most of the people I know will bathe themselves once a day, brush their teeth at least twice a day, and clean their clothes after wearing them a few times. However, it amazes me that people seldom clean their valuable jewelry even though they may wear it every day. What do you think would happen if you never washed your car? Not only would it look unsightly, but the dirt, oil, acid rain, and perhaps even road salt would shorten your vehicle's lifespan. The same thing is true for your gemstones. If a gemstone gets covered with dirt, body oils, perfumes, and abrasive cleansers, then serious damage to your jewelry can be the result.

Jewelry Boxes

After purchasing a piece of jewelry, a jewelry box is the next purchase you should make. Now that you know about the hardness, toughness, and stability of your gemstones, the importance of properly storing your jewelry should be obvious. Make sure that your box organizes your jewelry in such a way that your gemstones won't come in contact with each other. Don't forget—that diamond or sapphire will not only scratch many other gemstones, but can scratch your precious metals as well.

Cleaning Methods

The three basic ways to clean your jewelry are **ultrasonic cleaners**, **steam cleaners**, and **warm, soapy water**.

1. **Ultrasonic Cleaners**—sends high frequency sound waves through a liquid cleaning solution. Not safe for all gemstones, and over time may shake stones loose from their mountings.
2. **Steam Cleaners**—uses high pressure steam to quickly remove dirt and buildup. Not safe for all gemstones, particularly those which are heat sensitive, and also may loosen gemstones from their mountings.
3. **Warm, soapy water**—this is the method I recommend for all gems. For most gemstones, you can use a soft bristled toothbrush. On very soft gems such as pearls, you would not use a toothbrush. You could use a new make up brush or a lint free cloth.

I realize that many jewelers use ultrasonic or steam cleaners, but I don't recommend them for home use. Why? First of all, your jeweler knows which gems are safe to use in those cleaners, and they can also check the mountings afterward for any loose mountings.

There are gemstone cleaning solutions that are available and some are even safe for softer gemstones like pearls and opals. Always read the manufacturer's instructions carefully!

TREATMENTS

Some people are surprised to learn that their gemstone may have undergone a process to improve its appearance. This process is called a **treatment**. You might have heard it called an **enhancement**, but the Federal Trade Commission (FTC) says that treatment is a more accurate term. Should this alarm you? Absolutely not. Almost all gemstones are treated in one way or another. Actually, it is much rarer to find a gem that has <u>not</u> been treated than to find one that has. Consider this: whether you have spent $50,000.00 on a sapphire from the most elite jewelry store or $100.00 on a sapphire from a television shopping channel, most likely both sapphires have been heat treated. Your gemstone is no less valuable because it has been treated. In fact, your gemstone has become more desirable as a result of that treatment. Let us suppose that you are selling your home and you give it a treatment by painting the exterior and replacing the carpets. Do you think that treatment makes your home more desirable and valuable, or less? Obviously, this will make your home more desirable, and it should fetch a better price.

Certain treatments are more accepted than others, but all require disclosure by the seller. One thing that is important to keep in mind is that gemstones are as individual as people—no two are alike. You can't heat thousands of sapphires and expect all of them to suddenly become the ideal color. The potential for color must already exist in the gemstone. The treatment merely helps it along.

The following is a list of common gemstone treatments:

> **Bleaching**—a process to lighten or remove color. Most common in cultured pearls.
> **Cavity Filling**—a process to fill and seal voids to improve appearance and add weight. Sometimes used in rubies, sapphires, tourmaline, opals, and emeralds.
> **Colorless Impregnation**—using wax, plastic, or other substances to fill the pores and improve the appearance and stability of porous gemstones. Can be used in turquoise and jadeite.
> **Dyeing**—adding color to deepen, make color more even, or change color altogether. Commonly found in cultured pearls, jadeite, lapis lazuli, mother-of pearl, and turquoise.

Fracture Filling—filling narrow openings to improve the apparent clarity of a gemstone. Very common in emeralds.

Heat Treatment—this is the most common of all treatments. It can change the appearance of a gem by exposing it to rising temperatures. It can lighten, darken, deepen, or completely change a gemstone's color. It can also eliminate, create, or alter the appearance of inclusions as well. This treatment is considered stable for most gems. It is also likely the oldest of all gem treatments. In fact, examples of heat treated gems date back thousands of years. The list of gems that use this treatment are too numerous to mention.

Irradiation—changing a gemstone's color through electromagnetic radiation or bombardment with subatomic particles. This process has been around since the early 1900's. A few years back, I recall some news about fears associated with irradiated blue topaz. Today, irradiation is considered safe. In fact, use of atomic energy to irradiate gemstones is monitored by the Nuclear Regulatory Commission (NRC). Most common in topaz but can be used in other gemstones.

Smoke Treatment—a surface treatment used to darken an opal and improve its play of color. Low grade opal is wrapped in paper and roasted over fire. The sooty particles penetrate the porous opal and darken the background color.

Sugar Treatment—a surface treatment used to darken an opal and improve its play of color. Low grade opal is heated in a fruit juice solution saturated with sugar. After it cools and dries, it is immersed in sulfuric acid. The sugar converts to carbon and darkens the opal's color.

Surface Diffusion—a more controversial treatment that uses a combination of chemicals and high temperatures to create a shallow layer of color. The gem material is heated to near its melting point which allows the chemicals to penetrate the surface and become part of the crystal structure of the gemstone.

Surface Modifiers—treatments used to deepen color or give the appearance of colors on gems. These treatments, which include backing, coating, and painting have been around for thousands of years. These are the most superficial of all treatments and have no place in the legitimate gemstone world. However, they are still accepted in costume jewelry.

REAL versus FAKE

Have you ever heard terms used like **synthetic** and **simulated** and been confused by them? You're not alone. Many people are confused by these terms. In fact, even some people in the jewelry industry use these terms erroneously at best, and interchangeably at worst. This concept is much easier to understand than you think. But, before I explain what they mean, try not to think in terms of real versus fake. I only used these terms in the title to get your attention. Using the terms genuine versus not genuine can be just as confusing. Instead, let us think of a gem as being **natural**, **man-made**, or **imitation**.

Natural gemstones are created by nature. As I mentioned before, all gems have physical, chemical, and optical properties that make them unique. It can take many millions of years for Mother Nature to create a gemstone.

Synthetic gemstones have essentially the same chemical composition and structure as their natural counterparts but were created in a laboratory. The laboratory uses the same recipe that Mother Nature uses and tries to approximate conditions that exist on earth. The synthetic gemstone is created in a fraction of the time as one that is formed naturally. The terms "man-made" and "laboratory created" are describing the same thing. Is a synthetic gemstone genuine? You bet it is. It has the same properties as the natural stone, but it was created in a laboratory not mined from the earth. I think where everybody gets confused is how the word synthetic is used outside of the gemstone world. In the world of fashion, we think of synthetic fabrics like polyester as fake, and natural fibers like wool or cotton as real. Rest assured. A synthetic ruby is a genuine ruby, but it is not a natural ruby. Can we tell the difference between a natural gemstone and its synthetic counterpart? Usually, there are slight differences in the characteristics that allow a gemologist to distinguish between the two.

Simulated gemstones are the imitations. Simulated gemstones are stones that have the appearance of their natural counterparts. That is where the similarity ends. They do not have the chemical composition or structure of the natural gemstone. Green glass can be a simulated emerald, but it is not a synthetic emerald. Simulates are much more common in diamonds than in colored gemstones. Absolute™ and Diamonique™ are among the most famous examples of cubic zirconia used to simulate diamonds.

The most confusing simulated diamond seems to be moissanite. I have received many questions from people asking if moissanite is a synthetic diamond. It is not. There is such a thing as synthetic moissanite. The mineral moissanite was discovered in a meteorite in Diablo Canyon in Arizona. This mineral has also been created in a laboratory and is used to simulate a diamond. It does not have the same chemical properties as a diamond, it is merely a look alike.

To make matters just a bit more confusing, certain natural gemstones can be used to simulate other gemstones. An example would be using chromium diopside, a beautiful green gemstone, as a substitute for an emerald. Another would be using white topaz or white sapphire in a setting as an affordable alternative to using diamonds.

So why bother creating synthetic gems in a laboratory? For one thing, it is much more affordable to get a one carat synthetic emerald of great color and clarity than to find one of similar color and clarity in nature. Second, certain gems such as padparadscha sapphire are so exceedingly rare that shopping for a synthetic padparadscha sapphire may pretty much be your best chance of ever owning one.

PHENOMENAL GEMSTONES

No, I don't mean that they are really, really, really great. Phenomenal gemstones have striking optical effects that make them truly unique. The following list includes the most common properties of phenomenal gemstones:

1. **Adularescence**—the soft, delicate gleam of color that appears to float across a **moonstone**.
2. **Asterism**—also known as the star effect. This is caused by needle-like inclusions that are lined up in intersecting patterns. When they cross in the center, they create a star effect. These are usually found in **rubies**, **sapphires**, or **quartz**. Depending on the crystal structure of the gemstone, they may be a six rayed star or a four rayed star.
3. **Aventurescence**—sheen or glittery reflections off of small, flat inclusions within the gemstone. This can be found in **aventurine quartz** and **sunstone feldspar**.
4. **Chatoyancy**—this is also known as the **cat's eye** effect. This is caused by needle like inclusions that lie parallel to one another and reflect light in a narrow band. Most commonly found in **chrysoberyl** or **quartz**.
5. **Color change**—difference in the body color of a gemstone under different types of light. Most common in **alexandrite**, but also can occur in **sapphire**, **garnet**, or **spinel**.
6. **Iridescence**—a rainbow effect created when light is broken up into many colors. A non-gemstone example would be the colors you see on an oil slick or a soap bubble. Most commonly seen in **pearls** and **mother of pearl**, where it is called **orient**.
7. **Play of color**—patches of spectral colors as in an **opal**.

The Gemstone Chapters

Abalone shell is a highly underrated gemstone. It is not very expensive and, with its beautiful iridescence, can sometimes rival the beauty and colors of a black opal.

Abalone is mother-of-pearl that is from a mollusk class known as gastropada. Members of this class have one shell. They belong to a genus that scientists call "haliotis" which means "sea ear." This refers to the flattened shape of the shell, and it usually has a row of respiratory holes along one side not unlike the multiple pierce holes in many earlobes today.

Abalone
Paua Shell/Abalone Shell

The main sources of abalone today are Australia, New Zealand, and California. The abalone that comes from Australia and New Zealand is more often referred to as paua shell. Like its sister stones, pearl and mother-of-pearl, it is vulnerable to the ravages of pollution.

Most abalone shell is harvested for meat, but after the meat is consumed it is common to see the shell used in buttons and, frequently, in guitar inlay.

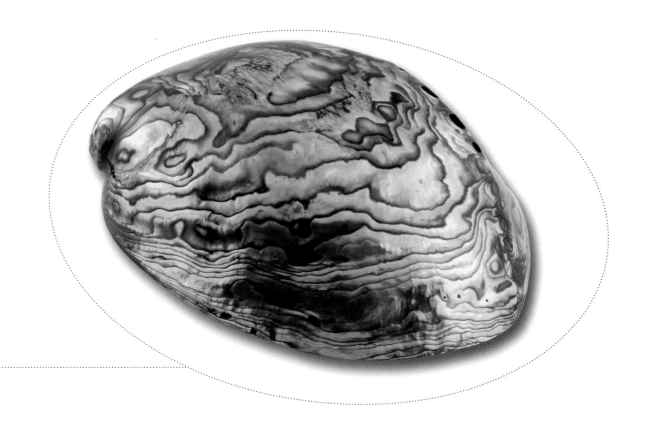

Composition—Calcium carbonate

Crystal Structure—Orthorhombic

Hardness—3.5   Toughness— Fair

Refractive Index—1.53 to 1.69   Specific Gravity—1.30

Stability: Heat—Burns if exposed to flame
               Light—Stable
               Chemicals—Easily damaged by acids

Treatments: Dyeing

Sources: Australia   New Zealand
             United States

Care and Cleaning: Steam Cleaning—Never
                   Ultrasonic—Never
                   Warm, soapy water—Safe

Agate is a form of chalcedony quartz that has been in use for thousands of years with the exception of fire agate which is believed to have been discovered sometime around the 1950s in Mexico. Most agates occur as nodules in volcanic lava. It is believed that water containing silica percolated through the rock and deposited a coating on the interior of openings within the lava rock. It is also interesting to note that petrified wood is actually wood that had its organic material replaced by agate. Agate can be found in a wide variety of colors, and its color is determined by different impurities that may be present during its formation.

Agate
Agate/Chalcedony

What sets agate apart from other types of chalcedony? Most agates are found in a variety of colors, and their distinct patterns appear as naturally occurring works of art. The varieties can look like moss, landscapes, trees, freeform geometric shapes, and bands that mirror the concentric circles of those found in tree trunks. In the case of fire agate, the stone can take on the appearance of the more expensive fine quality opals at a fraction of the cost.

Agate is usually cut into thin slabs or cabochons and then given a fine polish. It is generally not a faceted stone. Because of its toughness, it can also be carved into cameos or other interesting shapes. Larger pieces can even be used to create bookends, plates, and bowls.

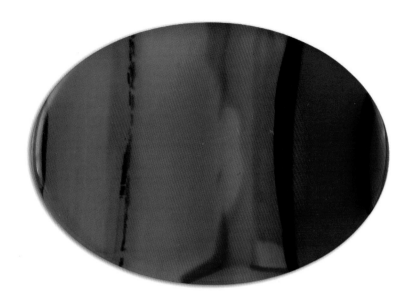

Composition—Silicone Dioxide

Crystal Structure—Trigonal

Hardness—7   Toughness—Good

Refractive Index—1.53-1.54   Specific Gravity—2.61

Stability: Heat—Color may change
               Light—Stable
               Chemicals—Avoid acids

Treatments: Dyeing
              Heating

Sources: Brazil   China
              Egypt   Germany
              India   Italy
              Mexico   Scotland
              United States   Uruguay

Care and Cleaning: Steam Cleaning—Not recommended
                   Ultrasonic—Not recommended
                   Warm, soapy water—Safe

Alexandrite is one of the rarest and most expensive gemstones in the world. It was named after the young Czar Alexander II, and was reportedly discovered on his coming of age birthday (the age of twelve) in 1830. Alexandrite is a phenomenal gemstone with amazing color change properties. The best quality alexandrite stones are a beautiful green color in daylight, or fluorescent light, and change to a red color in incandescent light. Ironically, red and green were also the colors of the Imperial Russian Flag. The original mines were located in the Ural Mountains of Russia, but very little, if any, is found there anymore. Large deposits were discovered in Brazil in 1987, but that supply has been largely depleted over the last few years. Small amounts are also mined in Sri Lanka and East Africa.

Alexandrite

Alexandrite/Chrysoberyl

Relatively recently, as of this writing, sizable deposits of alexandrite have been coming out of India. This has had a favorable impact on prices and availability. As a result, the television shopping channel HSN has been able to offer many pieces of jewelry of this once nearly impossible to find gemstone at relatively affordable prices.

Alexandrite remains an extremely rare gemstone. It is particularly difficult to find in larger carat weights. In fact, finding stones up to as little as one carat in size can prove to be quite a task. If you find you just can't live without large carat weights of alexandrite, you might want to look into the laboratory created variety. I have seen some great quality synthetic alexandrite from Chatham Created Gemstones at the Tucson Gem and Mineral Show the last few years, and it is a fraction of the cost of a comparable piece of natural alexandrite.

Anniversary stone—55th anniversary

Birthstone—June

Composition—Beryllium aluminum oxide

Crystal Structure—Orthorhombic

Hardness—8.5   Toughness—Excellent

Refractive Index—1.74 to 1.75   Specific Gravity—3.71

Stability: Heat—Stable
                Light—Stable
                Chemicals—Stable

Treatments: None

Sources: Brazil
                East Africa
                India
                Russia
                Sri Lanka

Care and Cleaning: Steam Cleaning—Safe
                        Ultrasonic—Safe
                        Warm, soapy water—Safe

Almandine is the most common and familiar of the garnet group of gemstones. If you purchase a piece of jewelry with a garnet gemstone, it is very likely that an almandine garnet was used in the setting. It is interesting to note that pure almandine is pretty rare. The true composition usually lies somewhere between almandine and pyrope, and these garnet stones get classified in whichever species its properties most closely resemble. Okay, maybe it isn't that interesting.

Almandine
Almandine/Almandite/Garnet

The name almandine comes from the Latin word "carbunculus alabandicus." "Carbunculus" means small coal, referring to the fire of the gem. "Alabandicus" refers to the city of Alabanda, an ancient city located in what is now Turkey, where the almandine gemstone was traded at that time.

For more information on the garnet group, or to learn about the folklore associated with this gem, please refer to the information listed under "Garnet."

Anniversary stone—2nd anniversary

Birthstone—January

Zodiac stone—Aquarius (January 21—February 21)

Composition—Iron aluminum silicate

Crystal Structure—Cubic

Hardness—7 to 7.5   Toughness—Fair to good

Refractive Index—1.76 to 1.83   Specific Gravity—4.00

Stability: Heat—Abrupt temperature changes may cause fracturing
              Light—Stable
              Chemicals—Stable, except concentrated hydrofluoric acid

Treatments: None

What causes the red color: Iron

Sources: Brazil   India
            Madagascar   Pakistan
            Sri Lanka   United States

Care and Cleaning: Steam Cleaning—Never
                   Ultrasonic—Usually safe
                   Warm, soapy water—Safe

Amazonite is a semi-opaque blue-green variety of microcline feldspar. Because it is opaque, it is usually seen in bead form or cut in cabochons. It can also be carved and used as cameos. Because of its color, it can be confused with turquoise or jade.

It is believed that amazonite is named after the Amazon River though it is not mined anywhere near there. It was more than likely confused with another similar stone that was mined near the Amazon river. There is another theory that suggests that since amazonite was widely used by the Egyptians, the name was derived from the legendary Amazon women warriors who referred to it as "the courage stone."

Amazonite
Amazonite/Microcline Feldspar

Composition—Potassium aluminum silicate

Crystal Structure—Triclinic

Hardness—6   Toughness—Good

Refractive Index—1.52 to 1.53   Specific Gravity—2.56

Stability: Heat—Stable
                Light—Stable
                Chemicals—Avoid acids

Treatments: Heat treatment

What causes the green color: Lead

Sources: Brazil   Canada
              India   Madagascar
              Namibia   Russia
              Tanzania   United States

Care and Cleaning: Steam Cleaning—Not recommended
                   Ultrasonic—Usually safe
                   Warm, soapy water—Safe

Amber, one of the organic gemstones, was created tens of millions of years ago when the sap of ancient trees hardened and fossilized. Anything that had the misfortune of getting trapped in that sap remains perfectly preserved today, even after millions of years. Amber is truly a window through which we can view history. It is best known for its golden amber color, and to this day is still sometimes referred to as "gold from the north." Other colors in which it can occur are white, orange, yellow, brown, green, blue, red, and even violet.

Amber

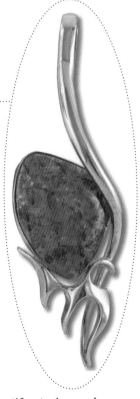

Amber has been popular for over 7,000 years. Many artifacts have been discovered that archeologists have dated back to 5000 BC. Few gemstones have been enjoyed longer by mankind. Approximately 600 BC, a Greek scientist discovered that amber had an amazing power. When rubbed with a cloth, it acquired the ability to attract feathers and other small objects. We now know that this "amazing power" is static electricity. In fact, the Greek word for amber is "electron."

Amber is often heated in oil to clarify it. These heat treatments can also produce disc-like stress fractures particularly in Baltic Amber. These are often referred to as "sun spangles," and can be quite beautiful.

Amber received a big surge in popularity after its starring role in the book and movie "Jurassic Park." In the story, blood was taken from a mosquito that was preserved in a piece of amber. Scientists then extracted dinosaur DNA from that blood to create a living dinosaur. While most scientists agree that recreating a life form from an incomplete DNA sample is highly unlikely, these specimens trapped in amber create a huge opportunity to collect a plethora of data on extinct plant and animal species.

Today, the most important sources of amber are the Baltic Sea coastal areas (Germany, Poland, and Russia) and the Dominican Republic. Much of the Baltic amber is found washing up on shore or floating on the water. Because the specific gravity is so low, it will float in salt water. Dominican amber is mined underground. Amber is also very low on the hardness scale—on average only 2 to 2.5. Because of this, mining is done by hand. Heavy machinery or blasting would cause too much damage to the amber. One interesting note—the longer the amber has been in the ground, the harder it will become. Amber from Myanmar (the oldest) has a hardness of about 3, while Dominican amber (younger than Myanmar amber and Baltic amber) is on the lower side of 2 on the hardness scale.

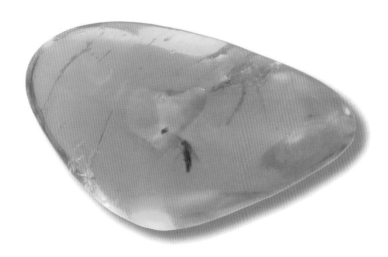

Composition—Mixture of organic plant resins

Crystal Structure—Non-crystalline

Hardness—2 to 2.5   Toughness—Poor

Refractive Index—1.54   Specific Gravity—1.05 to 1.09

Stability: Heat—Burns at low temperatures
                 Light—May darken with age
                 Chemicals—Avoid acids, caustics, alcohol, gasoline

Treatments: Heating
              Heating in oil
              Dyeing

Sources: The Baltic Coast—Germany
                                 Poland
                                 Russia
              Czech Republic   Mexico
              Canada   Myanmar (Burma)
              Dominican Republic   Romania
              France   Spain
              Italy   United States

Care and Cleaning: Steam Cleaning—Never
                   Ultrasonic—Never
                   Warm, soapy water—Safe, brushes not
                                       recommended

Amethyst is the most popular and valuable member of the quartz family of gemstones. It was once mined almost exclusively in Russia and was one of the most highly prized and expensive gemstones in the world. Indeed, at that time only royalty and the aristocracy could wear amethyst. Sometime, at about the turn of the twentieth century, large deposits of amethyst were discovered in South America and that made it more readily available to the average consumer. While the prices became much more affordable, the allure of amethyst has never waned.

Amethyst
Amethyst/Quartz

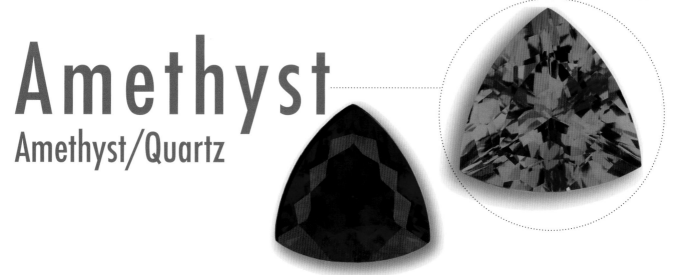

A legend says that Bacchus, the Roman god of wine, was so angry with mortals that he vowed to feed the next human to cross his path to the tigers. Unfortunately, a maiden by the name of Amethyst crossed his path on her way to worship Diana, the goddess of nature, fertility, and childbirth. Diana changed Amethyst into a pillar of colorless quartz to save her from the tigers. Bacchus was so impressed with Diana's miracle that he poured his wine over Amethyst changing her color to purple and the first amethyst gemstone was created. The name amethyst comes from a Greek word "amethustos," which means 'not drunken.' It was believed that wearing amethyst would protect you from drunkenness, although I have never successfully tested this premise. It is still advisable to not drink and drive no matter how much amethyst you might be wearing at the time.

Today, the major source for amethyst is Brazil where you can find a deep purple color and fairly large stones. Africa has also recently become an important source for amethyst. African amethyst is usually somewhat smaller than its Brazilian counterpart, but it is does have the most sought after color—a royal purple color with reddish overtones.

Anniversary stone—6th anniversary

Birthstone—February

Zodiac stone—Pisces (February 22—March 21)

Composition—Silicon dioxide

Crystal Structure—Trigonal

Hardness—7   Toughness—Good

Refractive Index—1.54 to 1.55   Specific Gravity—2.65

Stability: Heat—Extreme temperature changes can alter color or fracture
                    stone
                 Light—Some amethyst may fade
                 Chemicals—Avoid acids

Treatments: Heat treatment

What causes the purple color: Iron or manganese

Sources: Australia   Bolivia
                 Brazil   Canada
                 Germany   India
                 Madagascar   Mexico
                 Namibia   Russia
                 Sri Lanka   United States
                 Uruguay   Zambia

Care and Cleaning: Steam Cleaning—Not recommended
                             Ultrasonic—Usually safe
                             Warm, soapy water—Safe

Ametrine is a variety of quartz that has the colors of both amethyst and citrine in one stone. Theoretically, it can be found wherever amethyst or citrine is mined, but the only viable sources today are Brazil (where it was first discovered in 1979) and Bolivia. It is generally not a treated gemstone, although it is interesting to note that citrine can be created by heating amethyst, and amethyst can be created by irradiating citrine. This means that it is possible to create ametrine by heat and irradiation, but it is not done because natural stones are so affordable that it would not make economic sense to create stones by treating them.

Ametrine
Amethyst/Citrine/Quartz

Ametrine is not a birthstone in the literal sense, but it can be argued that it is a combination of the birthstone for both February and November.

Not a lot of folklore and legends for this one. After all, it was only discovered about 27 years ago. However, if you fancy a unique gemstone that shows two colors at once and is affordable and readily available, then ametrine is a great choice for you.

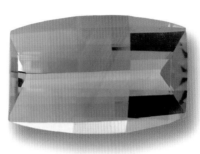

Composition—Silicon dioxide

Crystal Structure—Trigonal

Hardness—7   Toughness—Good

Refractive Index—1.54 to 1.55   Specific Gravity—2.65

Stability: Heat—Extreme temperature changes can alter color or fracture
                      stone
                      Light—Stable
                      Chemicals—Avoid acids

Treatments: None

Sources: Bolivia
                      Brazil

Care and Cleaning: Steam Cleaning—Not recommended
                      Ultrasonic—Usually safe
                      Warm, soapy water—Safe

Ammolite is a rare gemstone that was formed from the fossilized remains of the ammonite shell. Approximately 70 million years of pressure, heat, and mineralization resulted in the formation of the ammolite gemstone layer. While the fossil ammonite can be found in many locations on earth, the ammolite gemstone is currently only found in Southern Alberta, Canada. It was granted official gemstone status in 1981 by the International Commission of Coloured Gemstones. The word "rare" is admittedly overused when describing most gemstones, but in the case of ammolite it is certainly appropriate. At the current levels of production, it is believed that the supply of ammolite will be exhausted in about 20 years if no new sources are discovered.

Ammolite

Ammolite/Ammonite

Ammolite is a beautiful iridescent stone with a flash that is often compared to that of black opal. In fact, one of its nicknames is "opal of the sea." It is a stone that can be a bit pricier than some gemstones but far less than the price of a gem quality black opal.

You may notice that the hardness and toughness are not very high which might make you wary of using ammolite in jewelry. Under normal circumstances that would probably be true. However, most ammolite jewelry is set with triplet stones. The triplet has three layers—a base layer of shale, a middle layer of the ammolite gemstone, and a top layer of synthetic spinel which has a hardness of 8.5. This triplet construction makes it a much more durable stone for jewelry.

Ammolite was not commercially mined until about the mid 1970s, but its history dates back much further. It is believed that the ammolite was first discovered and used by the Blackfoot Indians who called it "Iniskim" or buffalo stone. It is said that one winter the Blackfoot were starving. The buffalo had disappeared and the snow was so deep that the tribe could not move to look for them. A woman was gathering firewood for the evening meal and the "spirit of the rock" came to her. She heard the stone say, "Woman, come and take me for I am powerful medicine. I am the greatest medicine of the buffalo." She took the stone back and the elders prepared a buffalo ceremony (I imagine they celebrated by drinking whatever the woman was drinking that made her hear rocks talking, but I digress). The next morning, a huge herd of buffalo was seen outside of their camp. Ever since, there has been a buffalo bundle with its Iniskim found in every painted lodge of the Blackfoot people.

Composition—Aragonite, Iron, Silica, Titanium, Copper

Crystal Structure—Orthorhombic

Hardness—4   Toughness—Fair

Refractive Index—1.52 to 1.67   Specific Gravity—2.8

Stability: Heat—High heat can affect epoxy and assembled stones
                Light—Stable
                Chemicals—Avoid acids

Treatments: Epoxy impregnation is used on some stones

Sources: Canada

Care and Cleaning: Steam Cleaning—Not recommended
                Ultrasonic—Not recommended
                Warm, soapy water—Safe, but do not soak
                 assembled stones for extended periods of time
                (doublets and triplets)

The name andalusite comes from Andalusia in Spain where it was first discovered. It is sometimes referred to as the poor man's alexandrite, because it appears to change color from a yellowish green to a reddish brown. This is not an accurate characterization since andalusite is not actually a color change stone. Alexandrite is a color change stone which means it will completely change colors depending on your light source, ie. indoor light vs. outdoor light. Conversely, andalusite is a strongly pleochroic stone which means you will see different colors when the stone is viewed in different directions.

Andalusite

That being said, andalusite is a very nice alternative to the rare alexandrite stone. Those of you with June birthdays who have been frustrated by the near impossible task of finding your alternative birthstone, the alexandrite, might want to consider andalusite.

Composition—Aluminum silicate

Crystal Structure—Orthorhombic

Hardness—7 to 7.5   Toughness—Fair to good

Refractive Index—1.63 to 1.64   Specific Gravity—3.16

Stability: Heat—Stable
               Light—Stable
               Chemicals—Avoid acids

Treatments: None

Sources: Australia   Brazil
              Canada   Myanmar (Burma)
              Russia   Spain
              Sri Lanka   United States

Care and Cleaning: Steam Cleaning—Not recommended
                    Ultrasonic—Usually safe
                    Warm, soapy water—Safe

Apatite is a beautiful gemstone that is comprised of calcium phosphate. It is the mineral that makes up the teeth and bones in all vertebrate animals. Apatite can deceive you by having the appearance of several other more common gemstones. In fact, the name comes from a Greek word "apatao," which means 'I am misleading.' It is most commonly found in a turquoise blue color, but can also be yellow, green, brown, white, and violet. It is also one of the few stones that might be a transparent gem faceted in a traditional manner or found as an opaque gem, cut in a cabochon, that can display a cat's eye effect.

Apatite

Apatite jewelry can be difficult to find. One of the reasons it is underutilized in jewelry is that it is a relatively soft gemstone, only 5 on the Mohs scale of hardness. Therefore, apatite is probably more suited for earrings, pendants, and brooches than for bracelets and rings. It is possible to enjoy an apatite ring, but I would recommend a protective setting such as a bezel setting over a prong setting.

Composition—Calcium Phosphate

Crystal Structure—Hexagonal

Hardness—5   Toughness—Fair

Refractive Index—1.63 to 1.64   Specific Gravity—3.20

Stability: Heat—Sensitive to extreme temperature changes
                Light—Stable
                Chemicals—Avoid acids

Treatments: Heat treatments

Sources: Africa   Brazil
             Canada   Germany
             Madagascar   Mexico
             Myanmar (Burma)   Russia
             Spain   Sri Lanka
             Sweden   Switzerland
             U.S.A.

Care and Cleaning: Steam Cleaning—Not recommended
                   Ultrasonic—Not recommended
                   Warm, soapy water—Safe

Aquamarine is a blue or blue green gemstone from the beryl family. Its blue color comes from trace amounts of iron. The name aquamarine comes from the Latin words for aqua and marina which means "sea water." It is difficult to believe that aquamarine and emerald come from the same gemstone family. While emerald is generally found in small sizes that can be heavily included, aquamarine is often mined in much larger crystals with excellent clarity.

Aquamarine is the birthstone for the month of March, although consumers have increasingly been substituting light shades of blue topaz as gemstones for their March birthdays. The reason for this can be summed up in one word—price. Blue topaz is a far more abundant and affordable stone than is aquamarine without having to sacrifice color, clarity, or durability.

Aquamarine
Aquamarine/Beryl

Legend says that King Neptune gave aquamarine as a gift to his mermaids. To this day, it is sometimes called the sailor's stone as carrying it is said to protect you while traveling at sea. Other folklore suggests that it can be a positive influence on marriages which makes it a great choice for an anniversary gift. Some even believed that wearing aquamarine would protect you from gossip, but I think that is just a lot of idle talk. Sorry, I couldn't resist.

Anniversary stone—19th Anniversary

Birthstone—March

Zodiac stone—Scorpio (October 24—November 21)

Composition—Beryllium aluminum silicate

Crystal Structure—Hexagonal

Hardness—7.5 to 8   Toughness—Good

Refractive Index—1.57 to 1.58   Specific Gravity—2.69

Stability: Heat—Exposure to heat not recommended
                 Light—Stable
                 Chemicals—Avoid acids

Treatments: Heating

What causes the blue color: Iron

Sources: Australia   Afghanistan
               Brazil   China
               India   Kenya
               Madagascar   Mozambique
               Nigeria   Pakistan
               Russia   United States
               Zambia

Care and Cleaning: Steam Cleaning—Usually safe
                   Ultrasonic—Usually safe
                   Warm, soapy water—Safe

Aventurine quartz is actually classified as a quartzite material. Technically, it is a rock not a mineral. What does this mean? It is made up of a combination of interlocking macrocrystalline quartz grains and grains of other color-imparting minerals. In the case of green aventurine, these color imparting minerals are a chromium rich form of mica called fuchsite which create the green color. These fuchsite crystals may also arrange themselves in such a way as to create a spangled effect known as aventurescence.

Aventurine
Aventurine/Quartz

Since a great deal of this material comes from India and has a similar appearance to jade, it is sometimes incorrectly called Indian jade. The name aventurine actually comes from the Italian phrase "a ventura" which means "by chance." This is an apparent reference to the accidental creation of the well known aventurine glass from the island of Murano in Venice, Italy. According to the story, a workman accidentally spilled copper filings into the molten glass mixture. The resulting beautiful effect became known as aventurino. This name then got passed on to the gem we know to day as aventurine, because it shares the same glittery appearance.

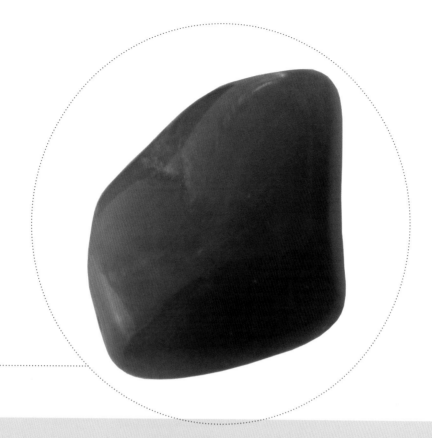

Composition—Silicone dioxide

Crystal Structure—Trigonal

Hardness—7   Toughness—Good

Refractive Index—1.54 to 1.55   Specific Gravity—2.65

Stability: Heat—Stable
              Light—Stable
              Chemicals—Avoid acids

Treatments: None

What causes the green color: Fuchsite mica

Sources: Brazil   India
             Japan   Russia
             Tanzania   United States

Care and Cleaning: Steam Cleaning—Not recommended
                   Ultrasonic—Usually safe
                   Warm, soapy water—Safe

Benitoite, like so many of today's rare and unusual gemstones, was discovered by accident. In 1906, a mining company sent prospectors to California's southern Mt. Diablo range to look for copper and mercury. Instead, they found a beautiful material they thought was sapphire. Further testing revealed that a new species of mineral had been discovered. It was named benitoite, after San Benito County in California where it is mined. In 1985 benitoite was officially named California's state gem.

Benitoite

Benitoite's two most distinguishing characteristics are the beautiful blue color, resembling sapphire or tanzanite, and the strong dispersion similar to a diamond. Wow, just think of the color of the most beautiful sapphire and the sparkle of a diamond! Okay, before we all get excited and run out looking for a benitoite ring, let's talk about the bad news (am I the life of the party or what?). It is extremely rare and pretty expensive. There is only one mine in the world, and it only operates about 6 to 8 weeks a year. Most of the material is very small ranging from 10 points to just over a carat. One of the largest stones is only 7.80 carats and is on display in the Smithsonian Institution in Washington D.C. The relatively soft stone (only 6.5 on the Mohs scale) is also brittle and not suitable for daily wear and tear, particularly in ring settings. So what is the bottom line? Benitoite is primarily a collector's stone, but in the gemstone business one never knows. Who knows if a new and more plentiful source will someday be discovered?

Composition—Barium titanium silicate

Crystal Structure—Hexagonal

Hardness—6.5   Toughness—Poor

Refractive Index—1.76 to 1.80   Specific Gravity—3.67

Stability: Heat—Exposure to heat not recommended
               Light—Stable
               Chemicals—Avoid acids

Treatments: None

Sources: United States

Care and Cleaning: Steam Cleaning—Not recommended
                   Ultrasonic—Not recommended
                   Warm, soapy water—Safe

You may not know the name beryl, but I suspect you have heard of its more familiar names -- emerald (colored by chromium and vanadium), green beryl (beryl colored by iron), aquamarine (blue beryl), morganite (pink beryl), heliodor (yellow beryl), goshenite (colorless beryl), and bixbite (red beryl). Beryl, in its purest form, would be colorless, but various trace elements indigenous to the area where it is mined will give it unique colors.

Beryl

There is a new form of pink beryl that comes from Madagascar called pezzottaite. It was named after Dr. Frederico Pezzotta, a well-known expert in Madagascar minerals. What sets this new pink beryl apart from morganite is that it contains cesium and has a much higher refractive index and specific gravity. Many of the pezzottaite specimens also have a chatoyant, or cat's eye effect.

I will go into much more detail in this book on the individual members of the beryl family under their more common names, but I wanted to mention something on beryl since you are still likely to hear names such as golden beryl or red beryl mentioned in the gem business.

Composition—Beryllium aluminum silicate

Crystal Structure—Hexagonal

Hardness—7.5 to 8   Toughness—Poor to good

Refractive Index—1.57 to 1.59   Specific Gravity—2.69 to 2.80

Stability: Heat—Exposure to heat not recommended
                Light—Stable
                Chemicals—Avoid acids

Treatments: Fracture filling
             Heating

Sources: Australia   Afghanistan   Brazil
             China   Columbia   Kenya
             Madagascar   Mozambique   Nigeria
             Pakistan   Russia   United States
             Zambia   Zimbabwe

Care and Cleaning: Steam Cleaning—Not recommended
                   Ultrasonic—Not recommended
                   Warm, soapy water—Safe

Bixbite is the rarest of all beryl gemstones. It is currently mined only in the Thomas Mountains and the Wah Wah Mountains of Utah. It has an intense red color that comes from the presence of manganese. It was named after the mineral collector Maynard Bixby. Gem quality material is extremely rare, and the largest faceted stones are less than 3 carats in size.

Bixbite

Bixbite/Beryl

Composition—Beryllium aluminum silicate

Crystal Structure—Hexagonal

Hardness—7.5 to 8   Toughness—Good

Refractive Index—1.58 to 1.59   Specific Gravity—2.80

Stability: Heat—Exposure to heat not recommended
                Light—Stable
                Chemicals—Avoid acids

Treatments: Usually none

What causes the red color: Manganese

Sources: United States

Care and Cleaning: Steam Cleaning—Not recommended
                   Ultrasonic—Not recommended
                   Warm, soapy water—Safe

Bloodstone is an opaque green chalcedony gemstone with red spots. The red spots come from the presence of iron oxide. These red spots give the appearance of drops of blood which is how bloodstone got its name. Early Christians believed that bloodstone was created when drops of blood from Christ's body fell on some green jasper at the foot of the cross. In medieval times, bloodstone was used to carve scenes of the crucifixion of martyrs which led to its also being called the martyr's stone.

Bloodstone
Bloodstone/Chalcedony

The Greek name for bloodstone is "heliotrope." The literal translation for this is "to turn toward the sun." Pliny the Elder, a Roman scholar and writer who is credited for writing the first book on gems (circa 70 A.D.), explains the name heliotrope as a depiction of the red reflection seen when a piece of bloodstone is placed in the water and pointed towards the sun. Perhaps it is a belief that the stone represents what the sun looks like as it lowers into the water at sunset.

Today, many people still believe in the mystical and healing properties of bloodstone. In countries such as India and Sri Lanka, it is ground into a fine powder and used as a medicine.

Bloodstone is less abundant than most of the other forms of chalcedony, but it is not a very expensive stone.

Alternate Birthstone—March

Zodiac stone—Aries (March 22—April 20)

Composition—Silicone Dioxide

Crystal Structure—Trigonal

Hardness—6.5 to 7   Toughness—Good

Refractive Index—1.53 to 1.54   Specific Gravity—2.61

Stability: Heat—Color may change
              Light—Stable
              Chemicals—Avoid acids

Treatments: None

Sources: Australia   Brazil
              China   India
              United States

Care and Cleaning: Steam Cleaning—Not recommended
                   Ultrasonic—Not recommended
                   Warm, soapy water—Safe

Carnelian is one of the most popular members of the chalcedony quartz family of gemstones. You may also see this gemstone referred to as cornelian. It is usually red or reddish orange in color which is probably how it got its name. It is believed that the word carnelian came from one of two origins: the Latin word "cornum," which means cherry, or "carnis," the Latin word for flesh.

Today, carnelian is usually found in bead form, in necklaces and bracelets, or in cabochon form for rings. It is rarely faceted. In the past, it was frequently carved into **cameo** (above the surface) or **intaglio** (below the surface) designs. Intaglio carved carnelian was particularly important in the days when correspondence was sealed with wax and then imprinted with the writer's mark. Wax does not easily adhere to polished carnelian, so it became popular in its use for rings used with sealing wax.

Carnelian
Carnelian/Chalcedony

Many myths surround the carnelian gemstone. The ancient Egyptians believed that the goddess Isis protected the dead by placing a carnelian amulet on the body of the deceased. They thought if the body was adorned in this fashion, Isis would protect them from harm as they moved through the afterlife. Carnelian is also one of the gemstones set into the breastplate of the high priest Aaron (Exodus 39:10-14) and represented one of the twelve tribes of Israel. Even today some believe that carnelian will help those who have bad memories, creative blocks, or confused minds (really Mom, that carnelian I gave you for your birthday was just a coincidence). It is also said to instill more confidence and a clearer speaking voice.

Zodiac stone—Virgo (August 23—September 22)

Composition—Silicone Dioxide

Crystal Structure—Trigonal

Hardness—6.5 to 7   Toughness—Good

Refractive Index—1.53 to 1.54   Specific Gravity—2.61

Stability: Heat—Color may change
                Light—Stable
                Chemicals—Avoid acids

Treatments: Heating

What causes the red color: Iron oxide

Sources: Brazil   India
              Uruguay   United States

Care and Cleaning: Steam Cleaning—Safe
                   Ultrasonic—Safe
                   Warm, soapy water—Safe

Chalcedony is an opaque gemstone that can occur in a wide range of colors. It is technically a member of the quartz family of gemstones. So why does it warrant a different name? When we think of the typical quartz gemstone, it occurs in large crystals like amethyst or citrine. Chalcedony gems are made up of many microscopic crystals that bond together to create one gemstone.

Chalcedony is one of the most common and least expensive gemstone materials on earth. That being said, some varieties (such as landscape agate) can be quite rare and very expensive. Chalcedony is found in many locations around the globe, far too numerous to name in this chapter. The name chalcedony is believed to be derived from Calcedon, an ancient port city on the Sea of Marmara (what is now Kadikoy, Turkey). Many of these microcrystalline quartz gemstones were mined and traded there.

Chalcedony
Microcrystalline Quartz

Chalcedony is most commonly set in jewelry in the form of cabachons. Typically, faceting an opaque gemstone does not have much of a desirable optical effect. However, some of the world's premiere jewelry designers have made a fortune setting chalcedony cabachons in their jewelry.

Like many gemstones that have been enjoyed for thousands of years, there is a lot of folklore associated with various varieties of chalcedony. It has been said that Greek sailors wore chalcedony for protection against drowning. In the 18th century, many Europeans believed chalcedony had the power to drive away evil spirits.

The following are some other types of chalcedony.

Varieties:
Agate	Amethystine	Bloodstone
Carnelian	Chrysocolla	Chrysoprase
Dendritic agate	Fire agate	Jasper
Landscape agate	Milky chalcedony	Moss agate
Onyx	Plasma	Sard
Sardonyx		

Composition—Silicon dioxide

Crystal Structure—Trigonal

Hardness—6.5 to 7   Toughness—Good

Refractive Index—1.53 to 1.54   Specific Gravity—2.61

Stability: Heat—Color may change in dyed material
              Light—Stable
              Chemicals—Avoid hydrofluoric and nitric acids

Treatments: Dyeing

Sources: Various locations throughout the world

Care and Cleaning: Steam Cleaning—Usually safe
                   Ultrasonic—Usually safe
                   Warm, soapy water—Safe

Diopside is a gem that can range from a bottle green color to nearly black in color. The green color usually comes from iron. However, when the green color comes from traces of chrome, it becomes a gorgeous emerald green color known as chromium diopside. It is distinctive in that it can have the color of the best quality emeralds and have great clarity as well. The downside is that it can be pretty expensive and is usually found in pretty small carat weights. It is pretty unusual to see many stones that are over one carat in weight. Ironically, in the rare occasion that you find a larger stone, the color is usually not as bright or vivid as in the smaller stones. This is because the trace element responsible for the beautiful green color also inhibits the growth of the crystal.

Chromium Diopside

Since most of the gem quality material comes from Siberia, there are also some mining challenges that contribute to the relative high cost of this gemstone. The hazardous conditions, particularly in the winter months, only allow for a few months of productive mining per year.

I must admit that, in my humble opinion, it is the most beautiful green gemstone available today.

Composition—Calcium magnesium silicate

Crystal Structure—Monoclinic

Hardness—5.5 to 6   Toughness—Good

Refractive Index—1.66 to 1.72   Specific Gravity—3.29

Stability: Heat—Stable
               Light—Stable
               Chemicals—Avoid harsh chemicals

Treatments: None

What causes the green color: Chromium

Sources: Myanmar (Burma)   Pakistan
             Siberia (Russia)   South Africa

Care and Cleaning: Steam Cleaning—Not recommended
                   Ultrasonic—Not recommended
                   Warm, soapy water—Safe

Chrysoprase is one of the most valuable members of the chalcedony family of gemstones. The name comes from the Greek words "chrysos" for golden and "prason" for leek. At that time, this name was used to describe any yellowish green gemstone. Today, it refers only to the green variety of microcrystalline quartz called chrysoprase.

Chrysoprase

Chrysoprase/Chalcedony/Microcrystalline Quartz

The best quality of chrysoprase is highly translucent and is apple green in color. Like most translucent gemstones, it is not faceted, is usually set in cabochon settings, or is carved. It is most often mistaken for fine quality jadeite jade. In fact, chrysoprase has sometimes been referred to by the misnomer of "Australian Imperial Jade." An estimated 85% of the gem quality chrysoprase comes from Queensland, Australia.

Composition—Silicon dioxide

Crystal Structure—Trigonal

Hardness—7   Toughness—Good

Refractive Index—1.53 to 1.54   Specific Gravity—2.61

Stability: Heat—Color may change in dyed material
              Light—Stable
              Chemicals—Avoid hydrofluoric and nitric acids

Treatments: Dyeing

What causes the green color: Nickel oxide compounds

Sources: Australia
             Brazil
             Russia
             United States

Care and Cleaning: Steam Cleaning—Usually safe
                   Ultrasonic—Usually safe
                   Warm, soapy water—Safe

Citrine is the yellow variety of quartz. The name is derived from the French word "citron," which means lemon. It can appear as a light yellow, a deeper orange yellow, or a reddish color that goes by the name of madeira citrine which was named for the color of wine. With most gems, you will find that the deeper the color the more valuable the stone. However, citrine seems to defy that premise. While the deeper colors are less abundant, the demand for, and prices of, citrine seem much more dependant on fashion trends. When fashion dictates that the brighter and lighter colors of yellow are in vogue, the prices for those corresponding colors of citrine will rise accordingly.

Citrine
Citrine/Quartz

How does it get its yellow color? Generally, it is the presence of iron, but very little citrine is found in that yellow color, at least not without some help. Most citrine starts out as either smoky quartz or amethyst and is heated to produce the yellow color. Usually, the light to medium yellows come from the smoky quartz and the deeper yellows and reds come from the amethyst.

Citrine is an affordable alternative to similar shades in topaz which is why it is now listed equally with topaz as a birthstone for November. It is also fairly easy to find citrine in larger carat weights. However, it should be noted that the citrine that comes from heating amethyst not only produces deeper colors, it typically yields smaller stones. It is common to find a specimen of light yellow citrine of 10 carats or more, but you are less likely to find the same sizes in Madeira citrine.

Anniversary stone—13th anniversary

Birthstone—November

Composition—Silicon dioxide

Crystal Structure—Trigonal

Hardness—7   Toughness—Good

Refractive Index—1.54 to 1.55   Specific Gravity—2.65

Stability: Heat—Extreme temperature changes can cause color loss or
                extreme changes can fracture stone
                Light—Stable
                Chemicals—Avoid acids

Treatments: Heat

What causes the yellow color: Iron

Sources: Bolivia   Brazil
             Madagascar   Russia
             Spain

Care and Cleaning: Steam Cleaning—Not recommended
                   Ultrasonic—Usually safe
                   Warm, soapy water—Safe

Coral is an organic gemstone which means it comes from a living thing. Specifically, it is the skeletons of tiny sea animals called coral polyps. Millions of these coral polyps join with others to form colonies. Each of them secretes a protective layer. The red, pink, white, and blue coral is comprised of pure calcium carbonate. The black and golden coral is made from a protein called conchiolin which forms the basic structure of all sea shells. These layers of calcium carbonate, or conchiolin become a home where the coral polyps hide when they are not feeding or if they feel they are in danger.

Coral

Coral is harvested in warm, tropical waters where the temperature is at least 68 degrees Fahrenheit. Coral cannot exist in cooler waters, so all of the current sources are located between 30 degrees north or 30 degrees south of the equator. It is mainly found in shallow waters, but some varieties have been known to grow at depths of up to 1000 feet.

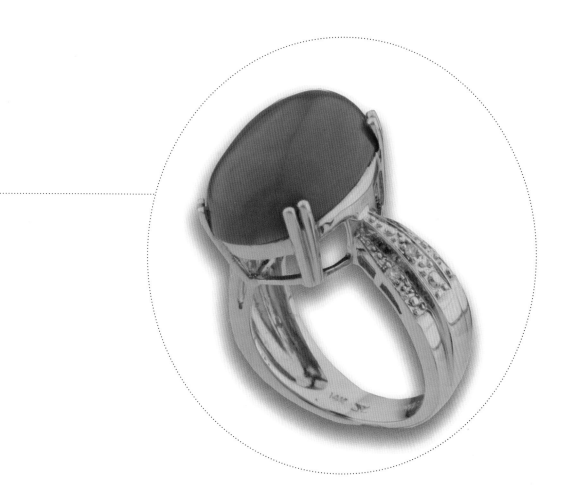

Coral can be found in red (sometimes called oxblood red), pink (called angel's skin pink), white, black, golden, and blue. The oxblood red variety of coral is generally regarded as the most valuable. At one time, the main source of this red coral was the Mediterranean Sea, but these waters have become so polluted that very little of this red material is still found. Efforts are now being made to try locating coral from greater depths in the Mediterranean Sea. Currently, most of the red coral comes from the Sea of Japan.

The main natural predator of coral is the Crown of Thorns starfish which hides among the colony and feeds on the coral polyps. However, the starfish is not the biggest threat to coral. Over-harvesting and pollution remain the most dangerous enemies of coral. As a result, there has been increasing international pressure for environmental efforts to protect coral reefs. Australia now prohibits the export of all native coral. Here in the United States, on December 4th, 2000, President Bill Clinton issued an executive order establishing the Northwestern Hawaiian Islands Coral Reef Ecosystem Reserve. The 84 million acre reserve is the largest protected area ever created in the United States.

Composition—Calcium carbonate

Crystal Structure—Trigonal

Hardness—3 to 4   Toughness— Fair to good

Refractive Index—1.49 to 1.66   Specific Gravity—2.68

Stability: Heat—Blackens or burns if exposed to flame
                Light—Stable
                Chemicals—Avoid acids, caustics,

Treatments: Bleaching
             Dyeing
             Impregnation with epoxy resins

Sources: Mediterranean Sea—Algeria, Morocco, Tunisia, Sardinia, Italy
                Pacific Ocean—Hawaii, Japan, Malaysia, Philippines, Taiwan
                Red Sea
                Persian Gulf

Care and Cleaning: Steam Cleaning—Never
                   Ultrasonic—Never
                   Warm, soapy water—Safe

Demantoid garnet is an extremely rare and valuable member of the andradite family of garnet gemstones. It has an emerald green color that comes from high levels of chromium, and has a higher dispersion than that of a diamond. Demantoid garnets frequently have a characteristic fingerprint called a "horsetail" which is caused by very fine, hairlike asbestos inclusions.

For more information on the garnet group, or to learn about the folklore associated with this gem, please refer to the information listed under "Garnet."

Demantoid
Demantoid/Andradite/Garnet

Anniversary stone—2nd anniversary

Birthstone—January

Zodiac stone—Aquarius (January 21—February 21)

Composition—Calcium iron silicate

Crystal Structure—Cubic

Hardness—6.5 to 7   Toughness—Fair to good

Refractive Index—1.85 to 1.89   Specific Gravity—3.85

Stability: Heat—Abrupt temperature changes may cause fracturing
                Light—Stable
                Chemicals—Avoid hydrofluoric acid

Treatments: None

What causes the green color: Titanium and manganese

Sources: Namibia   Russia
              Zaire

Care and Cleaning: Steam Cleaning—Never
                   Ultrasonic—Usually safe
                   Warm, soapy water—Safe

Drusy is the name given to gemstones where a layer of minute quartz stones have crystallized on the surface of another gemstone or mineral. It is the result of millions of years of water flowing over another stone and leaving silica deposits behind that eventually become tiny crystals on top of that stone. The result gives the appearance of the bright reflective surface of sugar or snow.

Drusy is a fascinating and beautiful gemstone that has become much more in vogue in the past ten years. Once thought of only as a collector's stone, a relatively new process puts a coating of titanium on the gemstone that creates spectacular and colorful results.

Drusy
Drusy Quartz

Most drusy that is sold is drusy quartz, but drusy can also occur in other minerals. The hardness and toughness is dependant on the material itself. In other words, drusy quartz would be a 7 on the hardness scale, and good on the toughness scale, but it is important to note that the toughness would not be as tough as a single crystal of the same gem. This is due to the nature of the tiny crystals that are attached to the original stone. Therefore, you not only have to think of the scratching or breaking aspect like typical gemstones, you have to worry about those tiny crystals detaching themselves from the host stone.

So what determines the value of drusy? We have already explained the four C's in this book, but when looking at a piece of drusy quartz, think of a fifth C: coverage. The evenness of the crystal coverage on top of the stone is an important factor, and affects the value of drusy gemstones.

Composition—Silicon dioxide

Crystal Structure—Trigonal

Hardness—7   Toughness—Good, but less than that
                                                                    of a single gem

Refractive Index—1.54 to 1.55   Specific Gravity—2.65

Stability: Heat—Extreme temperature changes can cause color loss or
                      extreme changes can fracture stone
                      Light—Stable
                      Chemicals—Avoid acids

Treatments: Some drusy stones are coated with titanium

What causes the vivid colors: Titanium coating

Sources: Many locations throughout the world; Brazil is most
                      prominent

Care and Cleaning: Steam Cleaning—Not recommended
                      Ultrasonic—Not recommended
                      Warm, soapy water—Safe

Emerald is the most valuable member of the beryl family of gemstones. Its color can range from a rich, deep green to a bluish green. The green color comes from traces of chromium and vanadium and, sometimes, minute traces of iron. There are green beryl stones that do not exhibit the typical emerald color or chemical composition. The green color of these stones comes from much heavier traces of iron, and these gems are simply called green beryl. The very definition of an emerald is that its green color must come from chromium. Green beryl does not contain chromium. One glance will show why these stones are not classified as emeralds.

Emerald

Emerald/Beryl

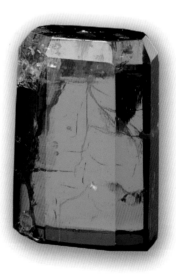

The history of emeralds dates back literally thousands of years. The name emerald comes from an ancient Greek word "smaragdus," which means green. From about 2000 B.C., the famed Cleopatra mines in Egypt were the world's only source of emeralds until sometime in the 1700s when the mines became so depleted that they finally closed. At about that same time, large deposits were discovered in Colombia which remains the most important source of emeralds today.

The infamous emperor Nero was said to have an extensive emerald collection and would stare at it for its calming effect. Legend says he also wore emerald glasses while watching the gladiators clash (Never mind we judge Nero too harshly for his eccentric behavior. Would any of us consider sitting in the Roman Coliseum for hours in the hot sun watching gladiators clash without our trusty sunglasses? I don't think so!). Emerald was also said to be one of the twelve stones used to adorn the breastplate of the high priest Aaron (Exodus 39) fashioned by Moses around the mid 1200s (B.C.). These twelve stones were chosen in the mid to late 1300s (B.C.) to represent the twelve tribes of Israel (Exodus 28). Emerald was chosen to represent the tribe of Levi.

While it is certainly no secret that gemstone mining can be a dangerous, if not lethal, business, perhaps nowhere is this more apparent than in the emerald mines of Colombia. As is often the case with any lucrative business, it can attract the unscrupulous and criminal element. In the late 1980s, Colombian drug lords were apparently dissatisfied merely running their profitable cocaine operations and decided to diversify into the booming emerald trade. As you might suspect, they simply chose to eliminate those who stood in their way. For many years, the Colombian government has tried to control the emerald mines as well as deter the drug traffic. The government has not had great success in either area. Needless to say, they have been fighting an uphill battle.

The three largest sources for emeralds are Colombia, Brazil, and Zambia. Zambian emeralds have been an increasingly popular choice as of late. Many people have asked me to explain the difference between Colombian and Zambian emeralds. Colombian emeralds still have the pedigree, color, and the mystique. Zambian emeralds seem to have fewer visible inclusions. To put it simply, Colombian emeralds get their color predominantly from chromium, so the color appears to be a deeper, truer green. Zambian emeralds have a higher concentration of the trace element vanadium, so they appear to be more bluish green in color. What these Zambian emeralds lack in color they more than make up for in clarity over their Colombian counterparts.

As discussed in the earlier chapter on clarity, every emerald has inclusions that make it unique much like a human fingerprint. The gemstone business is much more forgiving of these inclusions in an emerald than with any other gemstone. A relatively large number of inclusions is acceptable in an emerald and would not have much of a detrimental effect on its salability. If an amethyst had the same level of imperfections, it would have a very detrimental effect on its value.

Anniversary stone—20th and 35th Anniversaries

Birthstone—May

Zodiac stone—Cancer (June 22—July 22)

Composition—Beryllium aluminum silicate

Crystal Structure—Hexagonal

Hardness—7.5 to 8   Toughness—Poor to good

Refractive Index—1.57 to 1.58   Specific Gravity—2.71

Stability: Heat—Exposure to heat may cause fractures
                Light—Avoid intense light in fracture filled stones
                Chemicals—Avoid chemicals in fracture filled stones

Treatments: Fracture filling (also called oiling, may require retreating)
             Dyeing (rarely done, not permanent)
             Coating (rarely done, not permanent)

What causes the green color: Chromium and vanadium

Sources: Afghanistan   Australia
              Austria   Brazil
              Colombia   Egypt
              India   Norway
              Pakistan   Russia
              United States   Zambia
              Zimbabwe

Care and Cleaning: Steam Cleaning—Never
                   Ultrasonic—Never
                   Warm, soapy water—Safe

The garnet group is one of the most diverse and confusing of all gems. It includes almandine, pyrope, spessartite, tsavorite, hessonite, andradite (demantoid), rhodolite, uvarovite, malaya, and green grossular garnet. See what I mean about confusing? Each of these sub-species has unique properties that sets it apart from the rest. For example, almandine garnet will have different trace elements, a different refractive index, and a different specific gravity than pyrope garnet.

There is a lot of folklore associated with garnets. Throughout history they were thought to have special powers. Crusaders wore garnets for protection in battle and as a symbol of a safe return to their loved ones. King Solomon also wore garnet in times of battle.

Even as recently as 1892, tribal soldiers were said to have fired garnet bullets at the British army during the Kashmir Rebellion because it was believed that they were more fatal than lead bullets. Legend says that Noah used garnet to light his path and to illuminate the Ark. It was also said to be one of the twelve stones used to adorn the breastplate of the high priest Aaron (Exodus 39) fashioned by Moses around the mid 1200s (B.C.). These twelve stones were chosen in the mid to late 1300s (B.C.) to represent the twelve tribes of Israel (Exodus 28). Garnet was chosen to represent the tribe of Judah.

The name garnet comes from the Latin word "granatum" which means pomegranate seed. Garnet has been discovered in virtually all colors except blue. Of all these types, almandine (also referred to as almandite) is the variety most often seen when purchasing garnet jewelry.

The Garnet group:

Almandite (Almandine)
Andradite (Demantoid)
Grossularite (Hessonite, Tsavorite)
Malaya
Pyrope
Rhodolite
Spessartite
Uvarovite

Anniversary stone—2nd anniversary

Birthstone—January

Zodiac stone—Aquarius (January 21—February 21)

Composition—Varies, depending on the species

Crystal Structure—Cubic

Hardness—6.5 to 7.5   Toughness—Fair to good

Refractive Index—1.72 to 1.89   Specific Gravity—3.49 to 4.25

Stability: Heat—Varies depending on specific stone
               Light—Stable
               Chemicals—Avoid hydrofluoric acid

Treatments: None

Sources: Various locations throughout the world

Care and Cleaning: Check specific stones within garnet group

Goshenite is the colorless member of the beryl family of gemstones. It was named after the city of Goshen, Massachusetts where it was first discovered. It is not a widely known gemstone, although many years ago it was quite commonly used as a diamond substitute. It was also used on unsuspecting consumers as an emerald by backing it with green paint or foil and putting it in a closed setting where you could not see the back of the gem. Of course, this was before the days of the Federal Trade Commission, so such practices do not exist today where full disclosure is required in the gemstone industry.

Goshenite

Goshenite/White Aquamarine/Beryl

While goshenite is currently not a mainstream gemstone, it has been experiencing a surge in popularity as of late. One gemstone house has trademarked the name "white aquamarine" which certainly gives this gem a more exotic sounding name than goshenite. But, make no mistake about it, goshenite and white aquamarine are one and the same.

Composition—Beryllium aluminum silicate

Crystal Structure—Hexagonal

Hardness—7.5 to 8   Toughness—Good

Refractive Index—1.58 to 1.59   Specific Gravity—2.80

Stability: Heat—Exposure to heat not recommended
               Light—Stable
               Chemicals—Avoid acids

Sources: Brazil   Canada
               Russia

Care and Cleaning: Steam Cleaning—Usually safe
                   Ultrasonic—Usually safe
                   Warm, soapy water—Safe

Another member of the beryl family of gemstones is heliodor. Being a member of this distinguished family means it is related to the emerald, aquamarine, goshenite, morganite, bixbite, and golden beryl. Heliodor is the yellow form of beryl. The name translated comes from the Greek language and appropriately means "Gift of the Sun." What, you may ask, is the difference between golden beryl and heliodor? This is a valid question, and admittedly, there is a very fine line that separates them.

Heliodor

Heliodor/Yellow Beryl

I think the best way to describe the difference is to think of golden beryl as a true yellow or golden color reminiscent of citrine. Heliodor would be closer to a yellow with some greenish highlights like the hues of olive oil. Okay, not the most sophisticated explanation, but I think it gets the job done.

Heliodor is not a common stone, and fine specimens can command high prices among gemstone collectors.

Composition—Beryllium aluminum silicate

Crystal Structure—Hexagonal

Hardness—7.5   Toughness—Good

Refractive Index—1.57 to 1.58   Specific Gravity—2.80

Stability: Heat—Exposure to heat not recommended
                  Light—Stable
                  Chemicals—Avoid acids

Treatments: Some colorless beryl is irradiated to create heliodor

What causes the yellow color: Iron

Sources: Brazil   Madagascar
                Namibia

Care and Cleaning: Steam Cleaning—Usually safe
                   Ultrasonic—Usually safe
                   Warm, soapy water—Safe

Hematite is a beautiful and very affordable gemstone. It has a gray to black color with a high luster that can give it the appearance of a Tahitian black pearl. You need only pick it up to tell them apart. Hematite has the highest specific gravity of any natural gemstone. The hematite has such a high density (specific gravity of 5.20) that a hematite bead would feel about twice as heavy as a pearl of the same size. That is one of the reasons hematite is not used very often in earrings.

Hematite's chemical composition is iron oxide which is basically the same as rust. In fact, if you grind hematite into powdered form you get a red color. The name comes from the Greek word "haima," which means blood.

Hematite
Hematite

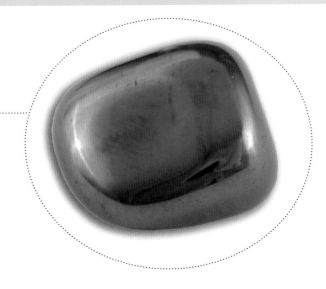

During Roman times hematite was associated with Mars, the god of war. It was believed to protect a warrior who rubbed it on his body. Native Americans also used hematite in its powdered form as a pigment for war paint.

Composition—Iron Oxide

Crystal Structure—Trigonal

Hardness—5.5 to 6.5   Toughness—Fair

Refractive Index—2.94 to 3.22   Specific Gravity—5.20

Stability: Heat—May become magnetic
               Light—Stable
               Chemicals—Soluble in hydrochloric acid

Treatments: None

Sources: Canada   England
               Norway   Sweden
               United States

Care and Cleaning: Steam Cleaning—Safe
                   Ultrasonic—Safe
                   Warm, soapy water—Safe

Grossular garnets can occur in a wide variety of colors. The green variety we know as tsavorite. When it is brownish yellow, brownish orange, or brownish red it is called hessonite. While this stone can have an absolutely beautiful appearance reminiscent of the color of cinnamon, you may be surprised to learn the name was derived from the Greek word "esson" which means inferior. Stones of that color were considered the least valuable at that time.

For more information on the garnet group, or to learn about the folklore associated with this gem, please refer to the information listed under "Garnet."

Hessonite

Hessonite/Grossularite/Garnet

Anniversary stone—2nd anniversary

Birthstone—January

Zodiac stone—Aquarius (January 21—February 21)

Composition—Calcium aluminum silicate

Crystal Structure—Cubic

Hardness—7 to 7.5   Toughness—Fair to good

Refractive Index—1.73 to 1.75   Specific Gravity—3.65

Stability: Heat—Abrupt temperature changes may cause fracturing
                Light—Stable
                Chemicals—Stable, except hydrofluoric acid

Treatments: None

What causes the orange color: Manganese and iron

Sources: Brazil   Canada
                Madagascar   Mexico
                Sri Lanka   Tanzania
                United States

Care and Cleaning: Steam Cleaning—Never
                   Ultrasonic—Usually safe
                   Warm, soapy water—Safe

Iolite may be the best kept secret in the gem business. In my opinion, it is the best value in a blue gemstone today. In fact, I have been saying it is one of the most underrated gemstones (not just blue) for about the past 17 years. Why am I so passionate about iolite? Consider that for the years 2001, 2002, 2003, and 2004, the top selling colored gemstone was sapphire. The number three or four best seller was tanzanite. You are probably thinking, "Okay, Paul, what is your point?" Here it is. Iolite is almost indistinguishable from the top color of tanzanite or sapphire, and it is a fraction of the cost of either! Like tanzanite, it is pleochochroic, which means that it shows different colors when viewed at different angles.

Iolite
Iolite/Cordierite

Unlike tanzanite and sapphire, it is much more readily available and is pretty easy to find in bigger carat weights. It is even a bit more durable than tanzanite. Incredibly, despite all of its assets, iolite has not gone mainstream as of yet.

The name iolite comes from the Greek words "ios," which means violet, and "lithos," which means stone. Iolite was used extensively by early Viking explorers. The sailors would use thin slivers of iolite to filter out haze allowing them to locate the exact position of the sun for navigational purposes. You might say they discovered the world's first polarized lens.

Iolite is also called cordierite or dichroite. Cordierite, the mineral from which we get iolite, was named after the French geologist Pierre Louis Cordier, who discovered it in 1809. Dichroism means essentially the same thing as pleochroism, but dichroism is a characteristic of a gem showing only two colors when viewed from different angles rather than more than two colors.

You may hear iolite also referred to as the water sapphire. It earned this misnomer because if you rotated a piece of iolite rough in your hand, sometimes it would look like a beautiful sapphire, and at other angles, it would look as clear as water. This phenomenon describes the great challenge for a gemstone cutter. They must exercise great skill when cutting to show the best shades of violet blue in the finished gemstone.

Anniversary stone—21ˢᵗ Anniversary

Composition—Magnesium aluminum silicate

Crystal Structure—Orthorhombic

Hardness—7 to 7.5   Toughness—Fair

Refractive Index—1.53 to 1.55   Specific Gravity—2.63

Stability: Heat—May damage stones
                  Light—Stable
                  Chemicals—Avoid acids

Treatments: None

What causes the blue color: Iron and magnesium

Sources: Brazil   Finland
                  India   Madagascar
                  Myanmar (Burma)   Namibia
                  Norway   Sri Lanka
                  Tanzania

Care and Cleaning: Steam Cleaning—Never
                               Ultrasonic—Never
                               Warm, soapy water—Safe

Ivory is one of six organic gemstones in the world, meaning it comes from living things. It is also one of the oldest known gemstones. Archaeologists in Europe have discovered ivory carvings and artifacts that are over 30,000 years old! While it seems that ivory has always been a part of our history, today it is one of the most controversial gemstones in the world.

Ivory

The United Nations Convention on the International Trade in Endangered Species of Wild Flora and Fauna prohibits trade in ivory. This means that it is illegal to buy or sell most ivory. As you may suspect, there are always those who are willing to circumvent the law at the expense of these endangered animals. To make matters more confusing, it is legal to trade in antique ivory (jewelry over 100 years old). However, let the buyer beware as some disreputable traders may use heat treatments to artificially make new ivory appear older.

In my opinion, the only safe bet to legally satisfy your need to own genuine ivory is to purchase fossil ivory. There is a decent supply of mammoth ivory that comes from the remains of the extinct wooly mammoth from areas such as Canada and Siberia.

Composition—Calcium hydroxyphosphate and organic

Crystal Structure—Amorphous

Hardness—2.25 to 2.75   Toughness—Fair

Refractive Index—1.53 to 1.54   Specific Gravity—1.90

Stability: Heat—May cause shrinkage, cracking, and discoloration
               Light—Yellows with age
               Chemicals—Vulnerable to many chemicals

Treatments: Bleaching, Dyeing, Heating

Sources: Elephant   Hippopotamus
               Narwhal   Sperm Whale
               Walrus   Wild Boar
               Dinosaur*   Mastodon*
               Wooly Mammoth*

               *Legal to trade in fossil ivory from these extinct species

Care and Cleaning: Steam Cleaning—Never
                   Ultrasonic—Never
                   Warm, soapy water—Safe

Jade has been utilized by many different cultures around the world for over 3,000 years. However, it was only in 1863 that we recognized that there are two distinctly different types of jade—jadeite and nephrite. While each has a unique hardness and chemical composition, they share one common trait. They are both the toughest gemstones in the world with a rating of exceptional on the toughness scale. The reason for this is that the structure of the gemstone is made up of interlocking grains of crystals.

Jade
Jadeite Jade

This tenacity makes jade very difficult to mine and to work with, but at the same time makes it nearly impossible to break or chip. This is why even relatively thin pieces of jade can be fashioned into many different shapes and intricate carvings. What does this means to those of you who love jewelry? It is a great every day gemstone to wear, it requires no special care, and it can withstand years of abuse.

The name jade comes from a Spanish term. When Cortez and his conquistadors were in Central America, they saw the indigenous people holding green stones against their sides believing that it could cure pain in their hips and kidneys. The conquistadors began to refer to this stone as "piedra de hijada" which means "stone of the loins." The French subsequently shortened this to "le jade" which is much closer to the word we use in English today.

Jadeite jade ranges from translucent to opaque. We always think of jade as being green, but it can occur in many other colors including black, brown, blue, lavender, orange, pink, red and white. Many of these other colors are achieved today by dyeing white stones, although they can occur naturally. Most of the green jadeite jade on the market today is not dyed.

The most sought after material is an emerald green nearly transparent stone referred to as "imperial jade." Imperial jade is extremely rare and very expensive. It is believed that centuries ago entire kingdoms were traded for a single piece of imperial jade. Today, jade is still very important to the Asian culture. It represents prosperity, success, and good fortune. While we in the western hemisphere think of precious gemstones as only diamonds, rubies, sapphires and emeralds, the Chinese consider jade the most precious of all gems.

Anniversary Stone—12th anniversary

Composition—Sodium aluminum silicate

Crystal Structure—Monoclinic

Hardness—7   Toughness—Exceptional

Refractive Index—1.66 to 1.68   Specific Gravity—3.33

Stability: Heat—May become damaged by a jeweler's torch
                Light—Stable
                Chemicals—Affected by warm acids

Treatments: Bleaching, Dyeing, Heating, Impregnation with paraffin wax

Sources: Guatemala   Japan
             Myanmar (Burma)   Russia
             Tibet   United States

Care and Cleaning: Steam Cleaning—Safe except for wax impregnated
                                   material
                   Ultrasonic—Safe except for wax impregnated material
                   Warm, soapy water—Safe

Nephrite jade has many of the same properties as the jadeite jade in the previous chapter. It can range from translucent to opaque; can appear from a light to dark green; and, can be found in black, brown, gray, yellow, and white. Once again, many of the colors of nephrite are achieved by dyeing the stones. Unlike jadeite, this dying often occurs in the green material. Most nephrite can be distinguished from jadeite by its darker almost greasy appearance.

The nephrite name came from the same story about the "stone of the loins," or "kidney stone" as it began to be called. The Latin translation for kidney stone is "lapis nephriticus" which later was shortened to nephrite.

Jade
Nephrite Jade

Anniversary Stone—12th anniversary

Composition—Calcium magnesium iron silicate

Crystal Structure—Monoclinic

Hardness—6 to 6.5   Toughness—Exceptional

Refractive Index—1.61 to 1.63   Specific Gravity—2.96

Stability: Heat—May become damaged by a jeweler's torch
                Light—Stable
                Chemicals—Affected by warm acids

Treatments: Bleaching, Dyeing, Heating, Impregnation with paraffin wax

Sources: Australia   Canada
             China   Mexico
             New Zealand   Russia
             Taiwan   United States

Care and Cleaning: Steam Cleaning—Safe except for wax impregnated
                                   material
                   Ultrasonic—Safe except for wax impregnated material
                   Warm, soapy water—Safe

There are not a lot of choices for jewelry buyers when it comes to pink gemstones. Looking for pink sapphires, pink diamonds, or even pink tourmaline can be a daunting task. If you are searching for any of the aforementioned stones in sizes larger than 1 carat, you should be prepared to pay a king's ransom. In 1902, renowned gemologist George Frederick Kunz identified a new pink gemstone variety of the mineral spodumene in California. He was so impressed with his find that he promptly named it after himself.

Kunzite
Kunzite/Spodumene

It should be noted that kunzite's color is a bit lighter and more pastel looking than many other pink gemstones, but it is much easier to find in larger carat weights and can be obtained at a much more reasonable price per carat.

So what, you may ask, is the downside of kunzite? First of all, it is not the most durable stone, rating poor on the toughness scale. Perhaps more importantly, this is a gemstone that is known to fade over time when exposed to intense light. That being said, this special care gemstone can truly be quite extraordinary if you choose a protective setting and avoid exposing your kunzite to direct sunlight for long periods of time. Some gemstone experts go so far as to recommend wearing kunzite only as an evening stone.

It is hard to imagine that a gemstone that has only been recognized for about 100 years would have much in the way of folklore and legends, but there is a somewhat amusing story about the origin of the name. George Frederick Kunz was a pioneer gemologist and a buyer for Tiffany and Company for many years. John Pierpont Morgan (J. P. Morgan) was a wealthy financier, gem enthusiast and an important customer of Tiffany and Company. Mr. Kunz promised his friend (and good customer) J. P. Morgan that he would honor him by naming his next gem discovery after him.

However, when the pink gemstone variety of spodumene was discovered, old George decided to name it kunzite, after himself. George made some excuse about not being able to reach J.P. Morgan from the west coast to obtain permission, and time was of the essence to name the stone, blah, blah, blah. This obviously did not sit too well with J. P. Morgan. All's well that ends well. A short time later (in 1911), pink beryl was discovered in Africa that Tiffany promoted as an alternative to kunzite and pink tourmaline. That new find, ironically very similar in appearance to kunzite, was named morganite, in honor of J. P. Morgan.

Composition—Lithium aluminum silicate

Crystal Structure—Monoclinic

Hardness—6.5 to 7   Toughness—Poor

Refractive Index—1.66 to 1.67   Specific Gravity—3.18

Stability: Heat—High heat can cause loss of color or fracture stone
              Light—Bright light can cause color to fade
              Chemicals—Can be attacked by concentrated hydrofluoric acid

Treatments: Irradiation

What causes the pink color: Manganese

Sources: Afghanistan   Brazil
             Madagascar   United States

Care and Cleaning: Steam Cleaning—Never
                   Ultrasonic—Never
                   Warm, soapy water—Safe

Labradorite is a phenomenal gemstone that displays an interesting play of color that is sometimes referred to as "**schiller**." This is caused by light reacting with many thin layers that make up the stone. Basically, the light goes in at one wavelength and slows down while traveling through the different materials inside the stone and comes out a different wavelength. This is what causes our eye to see different colors. The best way I can describe this phenomenon is to think of it as the combination of the play of color of an opal and the adularescence of a moonstone.

The gemstone was named after the main mining source which is Labrador, Canada.

Labradorite
Labradorite/Plagioclase Feldspar

Composition—Sodium calcium aluminosilicate

Crystal Structure—Triclinic

Hardness—6   Toughness—Fair

Refractive Index—1.56 to 1.57   Specific Gravity—2.70

Stability: Heat—Stable
               Light—Stable
               Chemicals—Avoid acids

Treatments: None

Sources: Finland
             India
             Labrador (Canada)
             Norway
             Russia

Care and Cleaning: Steam Cleaning—Not recommended
                   Ultrasonic—not recommended
                   Warm, soapy water—Safe

Lapis Lazuli is an opaque blue gemstone made up of lazurite, calcite, sodalite, pyrite and other minerals. The name comes from the Latin word "lapis lazulus" and the Arabic word "allazward" which translates to azure or azul. The word lapis means stone. The words azure and azul mean blue in French and Spanish respectively. So, Lapis Lazuli literally translates to blue stone. The ancient Romans called it "sapphirus," and indeed, quality lapis can look like the color of a fine sapphire.

Lapis Lazuli

The best quality lapis comes from Afghanistan and is known to have been mined there for over 6,000 years. Given the seemingly constant war torn nature of that country, the supply and price of lapis can be quite unpredictable. It will be interesting to see how the recent democratic reforms in Afghanistan will affect the worldwide lapis market. It remains to be seen as to whether years of warlords, terrorist camps, and opium drug trafficking used to support the brutal Taliban regime can make the transition to legitimate agriculture, mining and commerce in an open democratic society.

The most desirable color of lapis lazuli is the sapphire blue with no discoloration from the other minerals. It can also be found with white veining that comes from calcite or metallic diamond-like specks that are pieces of pyrite.

Lapis Lazuli was thought to be a cure for melancholy.

Anniversary stone—9th anniversary

Birthstone—December

Composition—Rock containing lazurite, calcite, pyrite and other minerals

Crystal Structure—Various

Hardness—5 to 6   Toughness—Fair

Refractive Index—1.50 to 1.67   Specific Gravity—2.80

Stability: Heat—High heat can affect color
                      Light—Stable
                      Chemicals—Decomposed by hydrochloric acid; discolored by
                                          cyanide solution

Treatments: Dyeing and impregnation

What causes the blue color: Lazurite

Sources: Afghanistan   Argentina
                      Chile   Russia

Care and Cleaning: Steam Cleaning—Never
                                 Ultrasonic—Never
                                 Warm, soapy water—Safe

Malachite is an opaque green gemstone usually found wherever copper is mined. Its history dates back to approximately 4000 B.C. where ancient Egyptians used malachite for jewelry. The name comes from the Greek word "malakhe" which means mallow, an apparent reference to an herb with dark green leaves.

As you would expect with a gem that has been mined for over 6,000 years, there are a lot of legends and folklore associated with malachite. It was believed that if you attached malachite to an infant's cradle, it would hold all evil spirits at bay and the child would sleep soundly and peacefully. It was also said that if you engraved the image of the sun on the surface of the stone, it would protect the wearer from enchantments, evil spirits, and from the attacks of venomous creatures.

Malachite

Malachite is not exactly a rare stone, but top grades are increasingly difficult to find. Over the years, copper miners would find great quantities of malachite during their tunneling but would simply crush and smelt it to recover the copper content within the rock. Recently, the demand for copper has skyrocketed due in large part to the industrial revolution in China. I suspect this increase in copper demand, and the resulting higher prices paid for it, may once again affect malachite supplies.

Malachite is a fairly soft stone (only 4 on the Mohs scale of hardness), but is pretty durable in the bead and cabochon form most often found in jewelry. It is also commonly found fashioned into dinner plates and goblets. The most desirable malachite has no pitting and might contain dramatic banding or floral patterns.

Composition—Copper Hydroxycarbonate

Crystal Structure—Monoclinic

Hardness—4   Toughness—Poor

Refractive Index—1.85   Specific Gravity—3.80

Stability: Heat—High heat may damage stone
               Light—Stable
               Chemicals—Avoid acids

Treatments: Impregnation with epoxy resin

What causes the green color: Copper

Sources: Australia   Russia
             United States   Zaire

Care and Cleaning: Steam Cleaning—Never
                   Ultrasonic—Never
                   Warm, soapy water—Safe

Malaya garnet is a beautiful orange gemstone that is similar in many respects to spessartite and hessonite garnets. Even the name sounds romantic, doesn't it? Well, you will be surprised to learn that the name malaya comes from a Swahili word that means "outcast" or "prostitute." This gemstone was so named because when it was first discovered in the 1970s, nobody wanted to buy it. Malaya has come a long way since. Today, fine quality malaya garnets can fetch a premium price. Malaya garnet is a hybrid of pyrope, almandine, and spessartite garnets.

Malaya

Malaya/ Garnet/Uvarovite

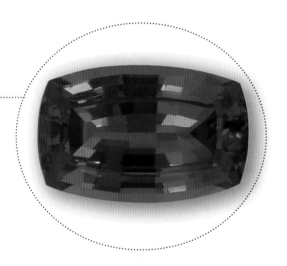

For more information on the garnet group, or to learn about the folklore associated with this gem, please refer to the information listed under "Garnet."

116

Anniversary stone—2nd anniversary

Birthstone—January

Zodiac stone—Aquarius (January 21—February 21)

Composition—Hybrid of pyrope, almandine, and spessartite garnets

Crystal Structure—Cubic

Hardness—7 to 7.5   Toughness—Fair to good

Refractive Index—1.77   Specific Gravity—3.80 to 4.16

Stability: Heat—Abrupt temperature changes may cause fracturing
                Light—Stable
                Chemicals—Stable

Treatments: None

What causes the orange color: Iron and chromium

Sources: Brazil   India
              Madagascar   Pakistan
              Sri Lanka   United States

Care and Cleaning: Steam Cleaning—Never
                   Ultrasonic—Usually safe
                   Warm, soapy water—Safe

Marcasite is a highly reflective gemstone that, at a distance, sparkles like a diamond. For many years, the Europeans referred to marcasite as the "black diamond." Most marcasite you see in jewelry today is actually a closely related mineral called pyrite. Most of us know yellow pyrite as "fool's gold." However, pyrite can also occur in white color, and this is what is used in the marcasite we see today.

Marcasite
Marcasite/Pyrite

The marcasite look first became popular during the Victorian Era. In fact, when Queen Victoria lost her husband there was a period of national mourning that lasted many months. During that time, it was customary to wear only black clothing (not that clothes were all that colorful in the 1800s to begin with). Since black was the fashion of the day, jewelry made of jet (a black gem made of fossilized coal) set in steel became very popular. These days, this look has evolved to sterling silver and marcasite.

Marcasite is a fairly brittle stone which can make it difficult to cut. Most marcasite stones are faceted on six sides and look somewhat like a diamond. In the late 1980s, the Swarovsky Company created a new way to facet marcasite on four sides, and it looks a little like a pyramid.

However, there are a few things worth thinking about concerning care and cleaning of your marcasite jewelry. First, many marcasite jewelry items are set in oxidized silver to make them look antique. Do not use solutions that remove tarnish from silver as that will remove this antique look. It is okay to store your marcasite jewelry in an anti-tarnish jewelry box. These boxes do not remove tarnish from jewelry, but they can prevent your jewelry from becoming tarnished.

Second, almost all marcasite is set in jewelry by using an adhesive rather than prongs or bezels. Because of this, do not immerse your marcasite jewelry in water or cleaning solutions for extended periods of time as this can break down the adhesives. Simply wiping down your marcasite, or a quick rinse, should always be sufficient.

I have always felt that marcasite represents one of the best values in jewelry today. It is quite inexpensive and can look better and more authentically antique the longer you own it.

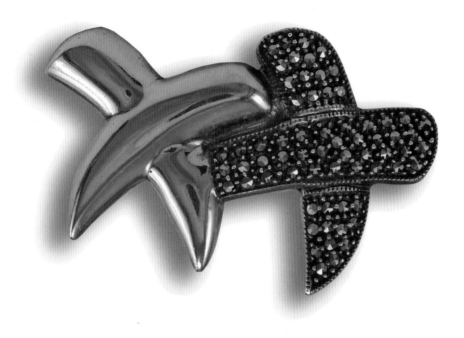

Composition—Iron sulphide

Crystal Structure—Cubic

Hardness—6 to 6.5   Toughness—Poor

Refractive Index—None   Specific Gravity—4.90

Stability: Heat—Stable
              Light—Stable
              Chemicals—Stable

Sources: France   Germany
              Italy   Mexico
              Norway   Peru
              Spain   Sweden
              Switzerland   United States

Care and Cleaning: Steam Cleaning—Usually safe
                   Ultrasonic—Usually safe
                   Warm, soapy water—Safe

Moldavite is a member of a family of gemstones called tektites. It is a naturally occurring transparent green glass. It was first discovered in the late 1700s near the Moldau River in what was then known as Czechoslovakia (now known as the Czech Republic). It was obviously named moldavite after the place in which it was found.

There are a couple of theories as to the origin of the stone. One school of thought is that it was carried here on a meteor that impacted the earth some 15 million years ago, its extraterrestrial contents melting into glass as it was burning up while entering the Earth's atmosphere. Another theory states that the meteor impacted the ground with such tremendous heat and pressure that it caused the rocks in the area to melt and cast them into the upper atmosphere before they fell back to the ground.

Moldavite
Moldavite/Tektite

For years, the Czech Republic was the only location where moldavite had been productively mined. However, recently I have seen samples of a beautiful green moldavite reported to have come out of China. We will have to see how that might impact the future availability and prices of moldavite.

Some legends suggest that the green stone associated with the Holy Grail was, in fact, a moldavite.

Composition—Silicon dioxide

Crystal Structure—Amorphous

Hardness—5.5 to 6.5   Toughness—Fair

Refractive Index—1.48 to 1.51   Specific Gravity—2.40

Stability: Heat—Stable
               Light—Stable
               Chemicals—Stable

Treatments: None

What causes the green color: Iron

Sources: China   Czech Republic

Care and Cleaning: Steam Cleaning—Not recommended
               Ultrasonic—Not Recommended
               Warm, soapy water—Safe

Moonstone is a unique gemstone that displays a phenomenon called "adularescence," which is a white or blue sheen that appears like a cloud moving across the stone when you rotate it in the light. This is sometimes referred to as "opalescence." Whatever you call it, this haunting moonlike phenomenon makes for a beautiful addition to any gemstone collector's repertoire.

Moonstone

Moonstone/Orthoclase Feldspar

The moonstone is believed to bring good fortune, and is regarded as a sacred stone in India. Some ancient people believed that moonstone could be used to see the future, but to do so required placing the stone in your mouth when the moon was full. Antoine Mizauld, a French physician and astrologer, wrote that moonstone could indicate the waxing and waning of the moon by a mark within the stone that would grow larger and smaller in relationship to the lunar cycle.

Composition—Potassium aluminum silicate

Crystal Structure—Monoclinic

Hardness—6 to 6.5   Toughness—Poor

Refractive Index—1.52 to 1.53   Specific Gravity—2.57

Stability: Heat—High heat can cause breakage.
                Light—Stable
                Chemicals—Attacked by hydrofluoric acids.

Treatments: None

Sources: India
                Myanmar (Burma)
                Sri Lanka

Care and Cleaning: Steam Cleaning—Never
                        Ultrasonic—Never
                        Warm, soapy water—Safe

Morganite is the pink member of the beryl family of gemstones making it a cousin to the emerald and the aquamarine. It was discovered in Madagascar in 1911, and was named after John Pierpont Morgan. J.P. Morgan was a banker, gem enthusiast and frequent customer of Tiffany and Company. George Frederick Kunz was a noted gemologist and vice president of Tiffany at the time. The story goes that George Kunz had promised to name his next gem discovery after his good friend (and valued customer) J.P. Morgan. However, when kunzite was discovered in California in 1902, George decided to name it after himself. He claims that he could not reach J.P. Morgan to get necessary approval for the trade name. Yeah, right George--and the dog ate my homework, and I ran out of gas--you get the idea. About 9 years later, George finally atoned for his sins by honoring Mr. Morgan after discovering the pink beryl in 1911.

Morganite
Morganite/Beryl

It would appear that J.P. Morgan got the last laugh. Not only is morganite very similar in appearance to the kunzite, it is also a more durable gemstone. It has a higher hardness rating, better toughness, and is more stable than George's kunzite.

Composition—Beryllium aluminum silicate

Crystal Structure—Hexagonal

Hardness—7.5 to 8	Toughness—Good
Refractive Index—1.58 to 1.59	Specific Gravity—2.80

Stability: Heat—High heat may cause pinkish orange stones to fade
                   to pink or cause fracturing.
           Light—Stable
           Chemicals—Resistant to all acids except hydrofluoric.

Treatments: Heat

What causes the pink color: Manganese

Sources: Afghanistan   Brazil
            Madagascar   United States

Care and Cleaning: Steam Cleaning—Not recommended
                        Ultrasonic—Not Recommended
                        Warm, soapy water—Safe

Mother-of-pearl is made from exactly the same material as that of pearls. A pearl is formed when a mollusk secretes a substance called nacre around an irritant inside the shell. That same substance lines the inside of the shell and is called mother-of-pearl. It is most commonly used as a face for many watches and is easy to identify with its pearl-like iridescent surface.

While mother-of-pearl is technically not a pearl, it can certainly be a beautiful piece of jewelry in its own right. It is usually extracted from the shell, carved into various shapes, and set into silver or gold settings. These days, it is more commonly being dyed various fashion colors.

For the purposes of this book, I am treating mother-of-pearl as an alternative to the pearl as far as birthstones and anniversary stones are concerned.

Mother-Of-Pearl

Anniversary stone—3rd anniversary, 30th anniversary

Birthstone—June

Composition—Calcium carbonate, conchiolin, and water

Crystal Structure—Orthorhombic

Hardness—2.5 to 4   Toughness— Good

Refractive Index—1.53 to 1.68   Specific Gravity—2.71

Stability: Heat—Burns if exposed to flame
               Light—Stable
               Chemicals—Avoid contact with all chemicals (hair spray,
                         cosmetics, and perfume can damage pearls)

Treatments: Bleaching
             Dyeing
             Irradiation

Sources: Natural Pearls—Persian Gulf, Gulf of Manaar (Indian Ocean),
                         Red Sea
             Cultured Pearls—Australia, China, Japan, and Polynesia
             Freshwater Cultured Pearls—China, Japan, United States

Care and Cleaning: Steam Cleaning—Never
                   Ultrasonic—Never
                   Warm, soapy water—Safe

Opal is one of the most fascinating gemstones in the world. William Shakespeare referred to opal as the "Queen of all gems," because it could exhibit the color of all of the world's gemstones in one single stone. Indeed, opal displays a phenomenon known as "play of color." This simply means the stone will show different flashes of color depending on the angle of viewing. These different color flashes are caused by tiny spheres of silica that bend the light into different spectral colors. The sizes of these individual spheres determine which colors the eye will see.

Opal

There are many different types of opal that you may encounter. In fact, they can be so strikingly different from one another that they all merit at least a sub-chapter in my hopefully upcoming, yet to be written, booklet on opals. Here is a little sampling: Black Opal, Boulder Opal, White or Crystal Opal, Jelly Opal, and Fire Opal just to name a few. Over the years, many people have asked me what my favorite gemstone is. Black opal is certainly at the top of a very short list. Not only is it strikingly beautiful, but it is extremely rare and can be prohibitively expensive.

One of the most common questions I receive about opals is, "I heard opals are bad luck. Is there any truth to this story?" Nothing could be further from the truth. Oh sure, there is the story about how Louis XI had the hands cut off of a goldsmith who broke one of his royal opals while setting it. Don't tell *him* opals aren't bad luck. The real rumor got started in the early 1800s when Sir Walter Scott wrote a novel titled "Anne of Geierstein." In this book, the main character spills holy water on an opal which subsequently begins to lose its color. As the color fades, so does the character's health leading to her ultimate death. People at that time incorrectly assumed that the author was trying to warn the of the bad luck of opals, and it literally destroyed the opal market for many decades that followed.

Here is one final comment about opals and current availability. An article in a recent issue of Colored Gemstone Magazine (July/August 2004) describes how opal is getting scarcer by the year. They estimate that production, in general, is down 70% from 10 years ago.

Anniversary stone—14th anniversary

Birthstone—October

Composition—Hydrated silica gel

Crystal Structure—Amorphous

Hardness—5 to 6.5   Toughness—Very poor to fair

Refractive Index—1.37 to 1.47   Specific Gravity—2.10

Stability: Heat—High heat or dramatic temperature changes can
                    cause cracking
               Light—Stable, but heat from intense lights can cause fracturing
               Chemicals—Avoid hydrofluoric acid and caustic alkalis

Treatments: Impregnation with oil, wax, or plastic
               Sugar treatment (soaking in sugar and acid)
               Smoke impregnation

What causes the play of color: Sub-microscopic spheres within the opal break
up light into spectral colors. The color the eye sees depends on the size of the
spheres.

Sources: Australia   Brazil
             Mexico   United States

Care and Cleaning: Steam Cleaning—Never
                   Ultrasonic—Never
                   Warm, soapy water—Safe

Pearls are one of the organic gemstones which means they come from living things. The others are amber, coral, ivory, jet, and shell. Pearls are also arguably one of the most versatile gemstones on the planet. They seem equally appropriate with your most elegant evening attire, your work wardrobe, or even the most casual vacation outfits.

Pearl

A pearl is formed when an irritant is introduced to a living mollusk. This can be naturally occurring, like a grain of sand, it can be a soft tissue or shell bead manually inserted by cultured pearl farmers. This irritant becomes the nucleus of the pearl. The shells of these mollusks are lined with a shiny lustrous substance called nacre. As a defense against this intruding irritant, the mollusk secretes many layers of this nacre to surround the irritant, and eventually a pearl is formed.

Pearls have a long and storied history. There is evidence that pearls have been worn and admired for over 5,500 years! I also like to think of pearls as the forgotten crown jewel. We always associate rubies, sapphires, emeralds and diamonds with royalty, but it is amazing how many pearls you see encrusting crowns, scepters, tiaras, and more in royal jewelry collections.

Pearls are evaluated differently than traditional colored gemstones. The four C's (cut, color, clarity, and carat weight) do not apply here. Instead, pearls are rated using the following criteria:

Luster—the amount of light reflected off of the surface of the pearl
Size—larger millimeter sizes are usually significantly more expensive
Shape—the best pearls are perfectly round or other symmetrical shapes
Surface—any imperfections visible on the surface of the pearl
Color—white or cream colors are still the most sought after, but fancy colors
          are increasing in popularity

It should be noted that the percentage of nacre can also have a huge impact on the value and longevity of your pearls. It takes anywhere from eighteen months to about three years for a pearl to be formed. Some cultured pearl farmers are introducing larger nuclei (which means it takes less nacre to form a pearl) or are harvesting the oysters in shorter periods of time which results in a thinner skin of the pearl.

There are many different types and names associated with pearls, so let's try to clear up some of the confusion. Here is a list of some of the classifications of pearls:

Natural—these form without any human intervention. Natural pearls are
          very rare and very expensive
Cultured—pearls that are farmed by inserting one or more nuclei into a
          mollusk
Akoya—saltwater cultured pearls from Japan and China
Blister pearls—cultured or natural pearls that form directly on the shell of
          the mollusk
Freshwater—pearls cultured in freshwater rather than salt water
Keshi—tiny pearls that form without being nucleated during cultured
          pearl cultivation
Mabe'—a cultured blister pearl assembled to a mother-of-pearl shell
Oriental pearls—natural pearls from the Persian Gulf
South Sea—larger saltwater cultured pearls (8mm to 18mm) from Australia,
          Indonesia, Myanmar (Burma), the Philippines, and Thailand
Tahitian—larger saltwater cultured pearls (8mm to 17 mm) from Tahiti
          and other French Polynesian Islands

Anniversary stone—3rd anniversary, 30th anniversary

Birthstone—June

Composition—Calcium carbonate, conchiolin, and water

Crystal Structure—Orthorhombic

Hardness—2.5 to 4   Toughness— Good

Refractive Index—1.53 to 1.68   Specific Gravity—2.71

Stability: Heat—Burns if exposed to flame
                Light—Stable
                Chemicals—Avoid contact with all chemicals (hair spray,
                cosmetics, and perfume can damage pearls)

Treatments: Bleaching
             Dyeing
             Irradiation

Sources: Natural Pearls—Persian Gulf, Gulf of Manaar (Indian Ocean),
                           Red Sea
             Cultured Pearls—Australia, China, Japan, and Polynesia
             Freshwater Cultured Pearls—China, Japan, United States

Care and Cleaning: Steam Cleaning—Never
                   Ultrasonic—Never
                   Warm, soapy water—Safe

Peridot is one of the most affordable transparent green gemstones in the world. It is believed that this gem was first mined about 1500 B.C. on an island off the coast of Egypt now known as "Zebirget" or "Zabargad." This island is also sometimes referred to as "St. John's Island." Are we confused yet? To make matters even worse, many years ago the island was called "Topazos," and the green gem mined there also was called Topazos, although the actual gemstone topaz has never been mined on this island.

Peridot
Peridot/Olivine

The Egyptians called peridot the "gem of the sun." Given the similar time frame of the Egyptian peridot mines (1500 B.C.) to the famed Egyptian Cleopatra emerald mines (2000 B.C.), it is easy to see why some jewelry historians believe that some (if not all) of the emeralds that Cleopatra wore may have, in fact, been peridot gemstones.

Today, the most important source for peridot is the San Carlos Indian Reservation in Arizona. It is estimated that as much as 80% of the world's peridot production comes from this one location. On the reservation, peridot can only be mined by individual Native Americans or by individual families of Native Americans from the San Carlos Reservation. While this rule protects both the environment and the Native Americans living on the reservation, it may also take longer for Arizona Peridot to become available to consumers. In fact, mining operations in the United States are subject to environmental, safety, and other laws that are not required in many overseas mining operations. This usually makes it more expensive to mine gemstones in the United States, and ultimately, makes the gems more costly to consumers. Because of this, China has made some tremendous inroads with some competitively priced peridot over the last few years.

Anniversary stone—16th anniversary

Birthstone—August

Zodiac stone—Libra (September 23—October 23)

Composition—Magnesium Iron Silicate

Crystal Structure—Orthorhombic

Hardness—6.5 to 7   Toughness—Fair to good

Refractive Index—1.64—1.69   Specific Gravity—3.34

Stability: Heat—Rapid heat changes can cause fracturing
              Light—Stable
              Chemicals—Avoid sulfuric and hydrochloric acids

Treatments: None

What causes the green color: Iron

Sources: Australia   Brazil
            China   Egypt
            Myanmar (Burma)   Norway
            Pakistan   South Africa
            United States

Care and Cleaning: Steam Cleaning—Never
                   Ultrasonic—Not recommended
                   Warm, soapy water—Safe

Prasiolite is a relatively new gemstone find. It has never been found in nature, but sometime around 1950 it was discovered that heating certain types of quartz gemstones from Brazil and Arizona in the presence of iron created a unique green color. The color can range from a pale yellow-green to a gorgeous seafoam green. Prasiolite is a relatively affordable gemstone, and it has been increasingly popular as of late. However, keeping in mind that not all quartz can be treated to create prasiolite, there is not an overly generous supply of this beautiful green stone available. Whether you are a collector of unusual gemstones, or you want a particular shade of green gem not easily found in any other stone, prasiolite is an excellent choice!

Prasiolite
Prasiolite/Vermarine/Quartz

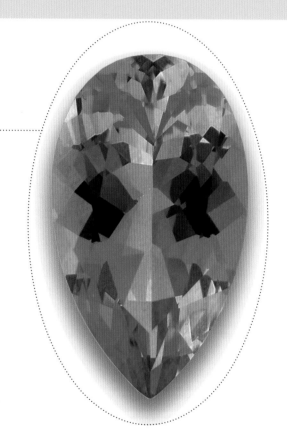

Composition—Silicon dioxide

Crystal Structure—Trigonal

Hardness—7   Toughness—Good

Refractive Index—1.54 to 1.55   Specific Gravity—2.65

Stability: Heat—Extreme temperature changes can alter color or fracture
                        stone
                   Light—Some amethyst may fade
                   Chemicals—Avoid acids

Treatments: Heat treatment

What causes the green color: Iron

Sources: Brazil
                   Arizona

Care and Cleaning: Steam Cleaning—Not recommended
                           Ultrasonic—Usually safe
                           Warm, soapy water—Safe

Pyrope is a beautiful blood red variety of garnet. In fact, this red gemstone can so closely resemble a ruby that at times you may hear it referred to as the "cape ruby." This is, of course, a misnomer. Pyrope is simply a magnesium rich variety of garnet that gets its bright red color from small quantities of chromium in the crystal structure. As discussed in the chapter on almandine, most garnets are somewhere between the chemical composition of an almandine and a pyrope garnet.

The name pyrope comes from the Greek word "pyropos," which means fiery.

For more information on the garnet group, or to learn about the folklore associated with this gem, please refer to the information listed under "Garnet."

Pyrope
Pyrope/Garnet

Anniversary stone—2nd anniversary

Birthstone—January

Zodiac stone—Aquarius (January 21—February 21)

Composition—Magnesium aluminum silicate

Crystal Structure—Cubic

Hardness—7.25   Toughness—Fair to good

Refractive Index—1.72 to 1.76   Specific Gravity—3.80

Stability: Heat—Abrupt temperature changes may cause fracturing
               Light—Stable
               Chemicals—Stable, except concentrated hydrofluoric acid

Treatments: None

What causes the red color: Chromium

Sources: Argentina   Australia
              Brazil   Mexico
              Myanmar   Scotland
              South Africa   Switzerland
              Tanzania   United States

Care and Cleaning: Steam Cleaning—Never
                   Ultrasonic—Usually safe
                   Warm, soapy water—Safe

Rhodochrosite is an opaque pink gemstone that can sometimes have white bands visible throughout the stone giving it a decorative effect similar to the rings of a tree. It is often confused with rhodonite, another opaque pink gemstone. Rhodonite usually has black veining instead of the white banding and is both harder and tougher than rhodochrosite.

The name rhodonite comes from the Greek work "rhodon" which means rose.

Rhodochrosite

Composition—Manganese carbonate

Crystal Structure—Trigonal

Hardness—4   Toughness—Poor to fair

Refractive Index—1.60 to 1.80   Specific Gravity—3.60

Stability: Heat—Stable
                Light—Stable
                Chemicals—Stable

Treatments: None

What causes the pink color: Manganese

Sources: Argentina   Mexico
                Namibia   Spain
                Russia   United States

Care and Cleaning: Steam Cleaning—Not recommended
                   Ultrasonic—Not recommended
                   Warm, soapy water—Safe

Rhodolite is a beautiful purplish red variety of garnet. Technically, it is a mixture of almandine and pyrope. I like to describe it as the best color of purple mixed with the best color of red. If amethyst and ruby ever had a child, it would be a rhodolite garnet.

Rhodolite was first discovered in North Carolina in 1898. It was called rhodolite to describe this new rhododendron color of garnet. The name rhodolite comes from the Greek words "rhodon," which means rose, and "lithos," which means stone. It should be noted here that stones being marketed as Brazilian plum garnet are actually rhodolite garnets from Brazil.

Rhodolite
Rhodolite/Garnet

For more information on the garnet group, or to learn about the folklore associated with this gem, please refer to the information listed under "Garnet."

Anniversary stone—2nd anniversary

Birthstone—January

Zodiac stone—Aquarius (January 21—February 21)

Composition—Magnesium iron aluminum silicate

Crystal Structure—Cubic

Hardness—7 to 7.5   Toughness—Fair to good

Refractive Index—1.76   Specific Gravity—3.74 to 3.94

Stability: Heat—Abrupt temperature changes may cause fracturing
                Light—Stable
                Chemicals—Stable, except hydrofluoric acid

Treatments: None

What causes the red color: Iron

Sources: Sri Lanka   Tanzania
                United States   Zimbabwe

Care and Cleaning: Steam Cleaning—Never
                Ultrasonic—Usually safe
                Warm, soapy water—Safe

Rhodonite is an opaque pink gemstone that can sometimes have black spots or veining throughout the stone. The black veining comes from oxidized manganese and adds a bit of visual interest and value to the stone. It is often confused with rhodochrosite, another opaque pink gemstone. The difference between the two is the hardness (rhodonite is 6 on the Mohs scale, rhodochrosite is only 4) and the white color banding visible in the rhodochrosite.

The name rhodonite comes from the Greek work "rhodon" which means rose.

Rhodonite

Composition—Manganese silicate

Crystal Structure—Triclinic

Hardness—6   Toughness—Good

Refractive Index—1.71 to 1.73   Specific Gravity—3.60

Stability: Heat—Stable
                     Light—Stable
                     Chemicals—Stable

Treatments: None

What causes the black veining: Oxidation of manganese

What causes the pink color: Manganese

Sources: Australia   India
                     Mexico   Sweden
                     Russia   United States

Care and Cleaning: Steam Cleaning—Not recommended
                              Ultrasonic—Not recommended
                              Warm, soapy water—Safe

Rose de France is a relatively recent discovery in the gemstone world. The reason for this is that it is technically just a lighter shade of amethyst that was given the trademarked name of "Rose de France." The story surrounding this is that a gentleman working for a prominent gemstone company in New York City was traveling in Paris, France. While visiting some gardens there, he noticed a lovely rose in a particular shade of lavender that he had never seen in a gemstone. He vowed then and there that should he ever discover a gem in that color, he would give it the name Rose de France after the unique color of rose he had seen on his trip.

Rose de France

Rose de France/Lavender Amethyst/Quartz

It is always amusing to me how these stories sometimes take on a life of their own. I have heard many variations of this story, and much like the game of telephone that we played as kids, the story seems to grow in fantastic ways with each telling. But, believe me, I know the gentleman who went to Paris and trademarked the name Rose de France. Those of you who have watched HSN over the years probably know him, too. His name is Larry Pereg, and he has appeared as a guest in gemstone shows on HSN many times over the years.

Composition—Silicon dioxide

Crystal Structure—Trigonal

Hardness—7   Toughness—Good

Refractive Index—1.54 to 1.55   Specific Gravity—2.65

Stability: Heat—Extreme temperature changes can alter color or fracture
                         stone
                       Light—May cause fading in some stones
                       Chemicals—Avoid acids

Treatments: Heat treatment

Sources: Australia   Bolivia
                       Brazil   Canada
                       Germany   India
                       Madagascar   Mexico
                       Namibia   Russia
                       Sri Lanka   United States
                       Uruguay   Zambia

Care and Cleaning: Steam Cleaning—Not recommended
                       Ultrasonic—Usually safe
                       Warm, soapy water—Safe

Rubies and sapphires are both members of the corundum family of gemstones. In fact, the only difference between the two is color. If corundum is red, it is a ruby. If it is any other color (including pink), it is a sapphire. Otherwise, they have exactly the same physical, chemical, and optical properties. Like sapphire, its sister stone, rubies are historically one of the most popular colored gemstones sold every year. Statistics show that rubies placed in the top five in colored gemstones sold in the United States for the years 2001 through 2004.

Ruby is a great every day gemstone to wear. With a hardness of 9 on the Mohs scale and a toughness rated excellent, it is one of the most durable gemstones on the planet.

Ruby
Ruby/Corundum

A lot of people have written to me asking about the best color for rubies. The best rubies are a medium to dark red with great clarity. Don't get caught up in whether it is called "pigeon blood red" or "Burmese" ruby. I always suggest common sense. If you like the red color, it has good clarity, and you can afford it, buy it!

The name ruby comes from the Latin word "ruber" which means red. In the ancient language of Sanskrit it is called "ratnaraj" which means the King of Gems. As you may suspect, this gem has a lot of folklore befitting a stone of such important stature. It is found in most crown jewel collections. Many ancient people believed that a stone the color of blood could make you invulnerable to wounds. Some even went so far as to imbed the rubies inside their bodies for protection in battle. With the way people are piercing their bodies and painting themselves with tattoos these days, maybe putting a ruby in your body doesn't sound that ridiculous to you.

In the present day, many people consider ruby to be the quintessential color of passion and romance. This theory may not be so far fetched. Many experts have concluded that the color red is stimulating while the color blue is calming. Ever notice how some restaurants have red tablecloths? It wasn't necessarily because those tablecloths were on sale. It stimulates the appetite.

The rubies from Myanmar (Burma) still have the pedigree, but most of the excitement as of this writing seems to be coming from Kenya and Madagascar. That being said, this is a gemstone that is becoming more and more difficult to locate in good color. If new sources are not found soon, the top quality rubies will continue to escalate in price.

Anniversary stone—15th anniversary, 40th anniversary, 80th anniversary

Birthstone—July

Zodiac stone—Capricorn (December 22—January 20)

Composition—Aluminum oxide

Crystal Structure—Trigonal

Hardness—9   Toughness—Excellent

Refractive Index—1.76 to 1.77   Specific Gravity—4.00

Stability: Heat—High heat can change color or clarity
                 Light—Stable
                 Chemicals—Generally stable, but always avoid harsh chemicals

Treatments: Heat
             Surface diffusion
             Fracture filling
             Cavity-filling with epoxy resin or glass

What causes the red color: Chromium

Sources: Australia   Afghanistan
               Kenya   Madagascar
               Myanmar (Burma)   Pakistan
               Sri Lanka   Tanzania
               Thailand   Vietnam

Care and Cleaning: Steam Cleaning—Usually safe
                   Ultrasonic—Usually safe
                   Warm, soapy water—Safe

Quartz may sometimes be found with inclusions that can create a spectacular appearance. Many different mineral species have been discovered growing within quartz crystals. One particular inclusion occasionally found in quartz is a mineral called rutile. Rutile is a major ore used in creating titanium. Titanium is a technologically advanced metal which is light weight yet very strong. When there are several of these rutile crystals within a quartz crystal, it can create the appearance of black or golden needles sprinkled throughout the gemstone. This exquisite work of art by nature is called rutilated quartz.

Rutilated quartz is by no means the most expensive of gemstones, but it is much rarer than the typical quartz stones such as amethyst and citrine, so be prepared to pay a little higher price for rutilated quartz.

Rutilated Quartz

Rutilated Quartz/Sagenite/Quartz

Composition—Silicon dioxide

Crystal Structure—Trigonal

Hardness—7   Toughness—Good

Refractive Index—1.54 to 1.55   Specific Gravity—2.65

Stability: Heat—Sudden temperature changes may cause fracturing, heating
                         may change color
                 Light—Stable
                 Chemicals—Avoid hydrofluoric acids and ammonium fluoride

Treatments: Heat treatment

Sources: Brazil   Germany
               India   Madagascar
               South Africa   Sri Lanka
               Switzerland

Care and Cleaning: Steam Cleaning—Never
                   Ultrasonic—Not recommended
                   Warm, soapy water—Safe

Sapphire is one of the most popular gemstones in the world. In fact, blue sapphire has been the top selling colored gemstone for many years. This should come as no surprise. Blue is the most popular color in the world. In fact, in 2003 four of the top ten selling gemstones were some shade of blue—Sapphire (#1), Tanzanite (#3), Blue Topaz (#6), and Aquamarine (#8).

Sapphire is also a very durable stone. With a rating of 9 on the Mohs scale, it is second only to a diamond in hardness, and is in a select group of gemstones rated just below jade in toughness. However, that does not mean that sapphires are indestructible. Despite their durability, sapphires (and all other gemstones) can become damaged and should be treated with respect and care.

Sapphire
Sapphire/Corundum

Sapphire is best known as the blue gemstone from the corundum family of gems. Truth is, sapphire can be any color except for red. Red corundum is a ruby. Other than that, rubies and sapphires are identical stones.

I believe that sapphire's grip on the number one position as the most popular selling colored gemstone is tenuous at best. This has nothing to do with sapphires dropping in popularity and everything to do with lack of supply of good quality blue stones. Don't get me wrong - For me, there is still nothing like a great blue sapphire. While supplies of good color and clarity of sapphires have dropped and prices have gone up, other blue stones (like blue topaz) have moved up the list. They are more readily available and are much more affordable.

Ancient cultures believed in many magical properties of gemstones. Many crown jewel collections have sapphires in them, and in crowns and tiaras, sapphires were often set higher than any other gemstone because it was the color closest to that of the heavens. But, here is a little known fun fact about sapphires you may not know. Some people centuries ago would grind up sapphires and mix them in their drink to help settle the stomach. Sounds silly, doesn't it? Well, it really isn't that funny when you consider that these primitive people unwittingly discovered the recipe for Maalox. The chemical composition for sapphires is aluminum oxide. The ingredient in Maalox that relieves your acid indigestion is aluminum hydroxide!

Anniversary stone—5th anniversary, 45th anniversary, 70th anniversary

Anniversary stone—5th anniversary, 45th anniversary, 70th anniversary

Birthstone—September

Zodiac stone—Taurus (April 21—May 21)

Composition—Aluminum oxide

Crystal Structure—Trigonal

Hardness—9   Toughness—Excellent

Refractive Index—1.76 to 1.77   Specific Gravity—4.00

Stability: Heat—High heat can change color
                Light—Stable
                Chemicals—Generally stable, but always avoid harsh chemicals

Treatments: Heat (occurs in about 95% of sapphires)
             Surface diffusion
             Fracture filling
             Cavity filling with epoxy resin or glass
             Irradiation

What causes the blue color: Iron and titanium

Sources: Australia   Cambodia
             China   India/Pakistan (Kashmir)
             Kenya   Madagascar
             Myanmar (Burma)   Nigeria
             Pakistan   Rwanda
             Sri Lanka   Tanzania
             Thailand   United States
             Vietnam

Care and Cleaning: Steam Cleaning—Usually safe
                   Ultrasonic—Usually safe
                   Warm, soapy water—Safe

Smoky quartz is the brown variety of quartz. It is one of the few gemstones sold in a brown color, and it is very affordable even in larger carat weights. It is a great gemstone choice when earthy colors like yellow, tan, and brown are in fashion.

This stone was once actively mined in the Cairngorm Mountains of Scotland, but these mines are no longer active. It is generally accepted that these mountains produced the finest quality stones ever mined. There must be something about the quartz in those high altitudes that creates conditions perfect for the formation of smoky quartz. Today, much of the smoky quartz is mined in the Swiss Alps and Brazil, although other sources are listed on the following page.

Smoky Quartz

Smoky Quartz/Quartz

It is believed that the brown color comes from natural irradiation within the ground, though these days science lends a hand. Many clear quartz stones are irradiated after they are mined in hopes of creating that unique brown color.

Some people still erroneously refer to this stone as "smoky topaz." This misnomer is misleading at best, and it is illegal and unethical to sell any quartz as topaz.

Composition—Silicon dioxide

Crystal Structure—Trigonal

Hardness—7   Toughness—Good

Refractive Index—1.54 to 1.55   Specific Gravity—2.65

Stability: Heat—Extreme temperature changes can cause color loss or
               extreme changes can fracture stone
               Light—Stable
               Chemicals—Avoid acids and ammonium fluoride

Treatments: Irradiation
             Heat

What causes the brown color: Irradiation within the ground

Sources: Australia   Brazil
             Madagascar   Spain
             Switzerland   United States

Care and Cleaning: Steam Cleaning—Not recommended
                   Ultrasonic—Usually safe
                   Warm, soapy water—Safe

Spessartite garnet is an uncommon member of the garnet family of gemstones and is very rare to find in gem quality. It has a bright orange or reddish orange appearance that comes from trace elements of iron in the garnet. The name spessartite comes from the Spessart district of Bavaria, Germany where it was first discovered.

For more information on the garnet group, or to learn about the folklore associated with this gem, please refer to the information listed under "Garnet."

Spessartite
Spessartite/Spessartine/Garnet

Anniversary stone—2nd anniversary

Birthstone—January

Zodiac stone—Aquarius (January 21—February 21)

Composition—Manganese aluminum silicate

Crystal Structure—Cubic

Hardness—7   Toughness—Fair to good

Refractive Index—1.79 to 1.81   Specific Gravity—4.16

Stability: Heat—Abrupt temperature changes may cause fracturing
                Light—Stable
                Chemicals—Stable, except concentrated hydrofluoric acid

Treatments: None

What causes the orange color: Iron

Sources: Brazil   India
             Madagascar   Pakistan
             Sri Lanka   United States

Care and Cleaning: Steam Cleaning—Never
                   Ultrasonic—Usually safe
                   Warm, soapy water—Safe

165

Spinel is a remarkable gemstone that very few people know. It is the Rodney Dangerfield of gemstones—it gets no respect! As I mentioned in chapter one of this book, I believe that the value of spinel would likely rise dramatically if only consumers could learn more about this stone. Spinel can occur in a wide variety of colors and is extremely durable. The different colors are caused by various trace elements, or impurities, within the stone.

Spinel

Spinel can often be found wherever rubies and sapphires are mined. It is also interesting to note how similar the physical, chemical, and optical properties are to those of corundum. Just how good a job does spinel do of imitating a ruby or sapphire? Consider that the Black Prince's Ruby, the large red gemstone found in the Imperial State Crown of England is, in fact, a fine red spinel. I am still mystified by the fact that fine gem quality red spinel is as rare as it is and still costs only a fraction of what a ruby of similar color and quality will command.

There are a couple of theories as to the origin of the name. It may be derived from the Latin word "spina" which means little thorn. This is in reference to the sharp triangular shape of some of the crystal faces. It could also come from the Greek word "spinter," which means spark, describing the fire from the red spinel in particular.

Anniversary Stone—22nd anniversary

Composition—Magnesium aluminum oxide

Crystal Structure—Cubic

Hardness—8	Toughness—Good
Refractive Index—1.71 to 1.73	Specific Gravity—3.60

Stability: Heat—High heat may cause fading
            Light—Stable
            Chemicals—Stable

Sources: Australia   Afghanistan
            Brazil   Cambodia
            Madagascar   Myanmar (Burma)
            Pakistan   Sri Lanka
            Russia   Tanzania
            Thailand   United States

Care and Cleaning: Steam Cleaning—Usually safe
                   Ultrasonic—Usually safe
                   Warm, soapy water—Safe

Sugilite is a relatively recent find in the gemstone world. It was first discovered in 1944 by Ken-ichi Sugi, a Japanese geologist, and this new gemstone was promptly given the name sugilite in his honor. There was not much material to be found, however, and the gem disappeared to relative obscurity until a large deposit was discovered in South Africa in 1975.

Sugilite is usually an opaque gemstone with color ranging from light lavender to deep purple. It is commonly found with black matrix inclusions. However, the best gem quality specimens are translucent, royal purple in color, and contain no visible black matrix.

Sugilite
Sugilite/Royal Lavulite/Royal Azel

Sugilite is sometimes referred to by the misnomer of purple turquoise. It actually has no relationship to the mineral turquoise whatsoever. However, like turquoise, it is commonly used in Southwestern and Native American jewelry designs.

Composition—Potassium Sodium Lithium Iron Manganese Aluminum
        Silicate

Crystal Structure—Hexagonal

Hardness—5.5 to 6.5   Toughness—Good

Refractive Index—1.59 to 1.61   Specific Gravity—2.68 to 2.79

Stability: Heat—Stable
                Light—Stable
                Chemicals—Avoid acids

Treatments: None

What causes the purple color: Manganese

Sources: Canada   India
              Japan   South Africa

Care and Cleaning: Steam Cleaning—Not recommended
                   Ultrasonic—Not recommended
                   Warm, soapy water—Safe

Sunstone is a member of the oligoclase feldspar family of gemstones which makes it a close relative of moonstone and labradorite. Some sunstone can have a ruby-like reddish color and virtually no visible inclusions. This pretty much describes the sunstone that comes from Oregon which is coincidentally their state gem. Fine gem specimens, like those I just described, command top prices in the gemstone market. Most sunstone has metallic inclusions giving it a sun-spangled glittery effect known as aventurescence. This is due to particles of iron trapped in the stone. This phenomenon can be quite attractive, though not quite as rare or pricey as the top grades of Oregon sunstone.

Sunstone

Sunstone/Oligoclase Feldspar

Composition—Sodium Calcium Aluminosilicate

Crystal Structure—Triclinic

Hardness—6 to 6.5   Toughness—Poor

Refractive Index—1.54 to 1.55   Specific Gravity—2.64

Stability: Heat—Stable
                     Light—Stable
                     Chemicals—Avoid acids

Treatments: None

What causes the yellow color: Iron

What causes the red color: Copper

What causes the metallic glitter: Inclusions of hematite (iron)

Sources: Canada   India
                     Norway   Russia
                     United States

Care and Cleaning: Steam Cleaning—Not recommended
                            Ultrasonic—Not recommended
                            Warm, soapy water—Safe

Tanzanite, the much sought after stone from east central Africa, is filled with so much beauty and so much romance. It also happens to be surrounded by so much misinformation! You will probably hear more falsehoods quoted about tanzanite than any other gemstone you encounter.

Stop me if you have heard some of these statements: "There is only one mine left," or "I heard the tanzanite mines are shutting down." One of my personal favorites is "The tanzanite mines are flooded, yada, yada, yada." This is utter nonsense. I have been to the mines in Tanzania, and they are not shutting down anytime soon. Somehow, the fact that there is only one geographic location where tanzanite is mined (Tanzania) got twisted into the statement that there is only one mine remaining. Don't feel bad, I have heard jewelry professionals quote the same inaccurate information.

Tanzanite
Tanzanite/Zoisite

Don't get me wrong. I am not saying that good quality tanzanite stones aren't rare. But, let's use some common sense. I defy you to go into any jewelry store in the United States and not find tanzanite on display. Tanzanite was also the third most popular colored gem sold in the year 2001; fourth in 2002; third again in 2003; and, fourth in 2004. Believe me, this could not happen if there was not an adequate supply.

Tanzanite is a relatively recent discovery in the gemstone world. It was discovered around 1962. The story goes that a couple of people in the gem business (from Germany) were looking in Tanzania for a new source for sapphires. They met a Masai tribesman who said he could take them to a place where there were beautiful blue gemstones. That stone turned out to be blue zoisite which we now know as tanzanite.

Around 1969, Tiffany and Company got wind of this discovery and started marketing this beautiful stone as the "poor man's sapphire." They named the gem tanzanite, after the country Tanzania, the only place in the world where it was mined. Obviously today, this gemstone has become very popular on its own merit. Its price can exceed that of many sapphires, so it is not a poor man's anything.

Tanzanite is a pleochroic stone which means it shows more than one color depending on how you orient the stone toward the light. If you hold up a piece of rough, you will generally see brown, blue, and purple. It is most commonly found in a predominantly brown color, and is heat treated to produce the blue and purple colors. The deeper blue colors are more expensive for two reasons. First, blue is a more popular color, so the demand is higher. Second, given the same amount of rough (raw gem material), a gem cutter can produce fewer blue stones from it than purple stones because of the way he must position the stone to cut it. This doesn't mean the purple stones are inferior, it just means you should pay less for them.

Tanzanite is a relatively soft stone, only 6 to 7 on the Mohs scale of hardness. Some jewelry professionals advise only wearing tanzanite in pendants or earrings. While I would consider tanzanite a special care stone, it is okay to wear tanzanite in rings. However, I would recommend a bezel or channel setting. These types of settings will better protect your tanzanite stones. If you treat your tanzanite with the proper care and respect, you should be able to enjoy it for years to come.

Anniversary stone—24th anniversary

Birthstone—December

Composition—Calcium aluminum hydroxysilicate

Crystal Structure—Orthorhombic

Hardness—6 to 7

Toughness—Fair to poor

Refractive Index—1.69 to 1.70

Specific Gravity—3.35

Stability: Heat—Abrupt temperature changes may cause fracturing
Light—Stable
Chemicals—Stable, except hydrofluoric and hydrochloric acids

Treatments: Heat

What causes the blue color: Vanadium

Sources: Kenya   Tanzania

Care and Cleaning: Steam Cleaning—Never
Ultrasonic—Never
Warm, soapy water—Safe

Tiger's-eye is one of the most affordable phenomenal gemstones in the world. The phenomenon of tiger's-eye is referred to as chatoyancy. Chatoyancy is the result of needle-like inclusions that are arranged in such a way that they create a straight line in the middle of a gemstone giving it the appearance of an eye. This effect is very similar to what a cat's eyes might look like when a flashlight catches its eyes in the dark. The origin of the name comes from this cat's-eye appearance. There are other types of chatoyant gemstones in quartz, but when the colors are those warm tones of yellow and brown, it is called tiger's-eye.

Tiger's-Eye
Tiger's-Eye/Quartz

Tiger's-eye is usually an opaque gemstone cut in cabochons and is primarily used in mens' jewelry.

Composition—Silicon dioxide

Crystal Structure—Trigonal

Hardness—7   Toughness—Good

Refractive Index—1.54 to 1.55   Specific Gravity—2.65

Stability: Heat—Sudden temperature changes may cause fracturing
               Light—Stable
               Chemicals—Avoid hydrofluoric acids and ammonium fluoride

Treatments: Bleaching
               Dyeing
               Heat treatment

Sources: Australia   India
               South Africa   Sri Lanka
               United States

Care and Cleaning: Steam Cleaning—Never
                   Ultrasonic—Usually safe
                   Warm, soapy water—Safe

Topaz is a gemstone that can occur in a wide range of colors (colorless, blue, yellow, pink, brown, green and red), but the majority of the topaz sold these days is blue. In fact, blue topaz consistently ranks in the top ten gems sold each year. Since it is so much more affordable than other blue stones such as sapphire, tanzanite, and aquamarine, it may continue to climb up that "top ten" ladder. Given the diminishing sources of sapphire in the world these days, perhaps blue topaz will eventually supplant sapphire as the world's most popular colored gemstone. It doesn't hurt that it is readily available in larger carat weights.

Topaz

Most blue topaz gemstones achieve their color through heat treatment, irradiation, or some combination of the two. Some people get very uneasy any time the word irradiation is used, but fear not—your skin will not glow like Three Mile Island after wearing your blue topaz. In fact, the Nuclear Regulatory Commission closely monitors and regulates all of the blue topaz that is treated with irradiation and will not release the stones until they are safe. This quarantine lasts anywhere from 3 months to a year.

Another interesting development in topaz lately is the increase in popularity of optically coated stones. You may have heard of these referred to as "mystic topaz" or "rainbow topaz" to name but a few of the trademarked terms. These stones receive a chemical coating (like a thin layer of film) over the stone that gives them a spectacular optical effect. It should be noted that this is not considered a permanent treatment by the Gemological Institute of America because this treatment is only skin deep. However, these coated treatments have had a dramatic impact on the price and availability of blue topaz as of late. The demand for colorless topaz stones has increased which then decreases the availability of stones that could become the blue topaz you know and love.

The name topaz comes from the Sanskrit word "tapas" which means fire. It is also believed to be named for an island in the Red Sea the Greeks called "Topazios" (currently named Zabargad) because of the green gemstone that was mined there. Ironically, the gemstone mined and traded there was not topaz at all. It was, in fact, peridot.

Anniversary stone—4th anniversary, 23rd anniversary

Birthstone—November (yellow), December (blue)

Zodiac stone—Sagittarius (November 22—December 21)

Composition—Aluminum fluorohydroxysilicate

Crystal Structure—Orthorhombic

Hardness—8   Toughness—Poor

Refractive Index—1.62 to 1.63   Specific Gravity—3.54

Stability: Heat—High heat can change color; sudden temperature change
                  may cause breaks
               Light—Stable
               Chemicals—Very slightly affected

Treatments: Heat or irradiation

Sources: Australia   Brazil
             Madagascar   Mexico
             Myanmar (Burma)   Namibia
             Nigeria   Pakistan
             Sri Lanka   United States

Care and Cleaning: Steam Cleaning—Never
                   Ultrasonic—Never
                   Warm, soapy water—Safe

Tourmaline is available in such a wide variety of beautiful colors that the name is derived from the Sinhalese word "toramalli" which means mixed precious stones. Before you laugh at the expense of the people of Sri Lanka, consider that we, who speak English, have even further confused the issue by applying so many different names to all the colors of tourmaline. Here is a list of the more common names for tourmaline:

Tourmaline

Rubellite—red tourmaline
Indicolite—violet to greenish blue tourmaline
Dravite—yellow to brown tourmaline
Achroite—colorless tourmaline
Watermelon tourmaline—green and pink tourmaline
Paraiba tourmaline—blue tourmaline from Paraiba, Brazil

Tourmaline is a beautiful gemstone that is also quite durable. The supply and demand seems to fluctuate wildly for many different reasons. Whenever the green and pink colors are hot in fashions, the demand seems to increase for green and pink tourmaline. In 2001, both green and pink tourmaline were in the top ten of the best selling gemstones for that year. They haven't been in the top ten since. I also recall when you could only get pink tourmaline out of Nigeria, and it was pretty pricey. Then, large deposits were discovered in Brazil which lowered the prices worldwide for pink tourmaline.

The prices for tourmaline can also vary greatly depending on the color. Intense reds, like rubellite or the rare parti-colored tourmalines showing more than one color, can command premium prices. The relatively new paraiba tourmaline was discovered in about 1988 (in Paraiba, Brazil), and is a magnificent blue color. It is one of the more rare and costly gemstones on the planet fetching prices in the thousands of dollars per carat.

Anniversary stone—8th anniversary (green tourmaline)

Birthstone—October

Composition—Complex borosilicate

Crystal Structure—Trigonal

Hardness—7 to 7.5   Toughness—Fair

Refractive Index—1.62 to 1.64   Specific Gravity—3.06

Stability: Heat—High heat can change color; sudden temperature change
                    may cause breaks
                Light—Stable
                Chemicals—Stable

Treatments: Heat, irradiation or acid

Sources: Afghanistan   Brazil
             Kenya   Madagascar
             Mozambique   Myanmar (Burma)
             Namibia   Pakistan
             Russia   South Africa
             Sri Lanka   United States

Care and Cleaning: Steam Cleaning—Not recommended
                   Ultrasonic—Not recommended
                   Warm, soapy water—Safe

Tsavorite garnet is one of the most intriguing members of the garnet family of gemstones. It is a bright green transparent crystal variety of green grossular garnet. It is predominantly mined in Kenya and Tanzania in east central Africa. The name "grossular" comes from the botanical name of the gooseberry which is "grossularia." The name "tsavorite" comes from Tsavo National Park in Kenya where it was first mined. This name was selected by Tiffany and Company when they first introduced the gemstone to the United States market in the early 1970s.

Tsavorite
Tsavorite/Grossularite/Garnet

While tsavorite is a garnet and a legitimate birthstone choice for those born in January, it is sometimes used as an exotic substitute for an emerald. It is a clearer and more durable gemstone than the emerald, but is a different shade of green. Before everybody gets all excited and runs out to stock up on tsavorite, it should be noted that it is a relatively rare and expensive stone particularly for a garnet. It is also generally found in smaller sizes. Most tsavorite used in jewelry is less than one carat in size. If you are fortunate enough to find it in larger sizes, be prepared to pay a premium price.

For more information on the garnet group, or to learn about the folklore associated with this gem, please refer to the information listed under "Garnet."

Anniversary stone—2nd anniversary

Birthstone—January

Zodiac stone—Aquarius (January 21—February 21)

Composition—Calcium aluminum silicate

Crystal Structure—Cubic

Hardness—7 to 7.5   Toughness—Fair to good

Refractive Index—1.69 to 1.73   Specific Gravity—3.49

Stability: Heat—Abrupt temperature changes may cause fracturing
               Light—Stable
               Chemicals—Stable, except hydrofluoric acid

Treatments: None

What causes the green color: Chromium and vanadium

Sources: Kenya   Tanzania

Care and Cleaning: Steam Cleaning—Never
                         Ultrasonic—Usually safe
                         Warm, soapy water—Safe

Turquoise is one of the most historically significant gemstones. There is evidence that ancient Egyptians were working with turquoise for jewelry over 7,500 years ago! It is believed that the name came from the fact that the stone was first introduced to Europe by traders from Turkey. These traders were called "Turks," and the stone they traded was called the "Turkish stone." The spelling we use today was probably derived from the French phrase "pierre tourques" which means Turkish stone.

Turquoise

Turquoise was actively mined as far back as 2100 B.C. in Persia (the area now known as Iran). The turquoise that was mined there is referred to as Persian turquoise, and that name still carries a certain pedigree. Today, the largest producers of turquoise are the United States (Arizona, New Mexico and Nevada) and China. It is also mined to a lesser extent in Mexico, Afghanistan, Chile, Russia, Australia, and Iran.

Turquoise is usually an opaque blue gemstone, although it is also common to see some green colors of turquoise. The best quality stones are referred to as robin's egg blue and have no evidence of veining or spider web matrix on the surface.

Turquoise is often associated with Native American jewelry and culture. Mining and manufacturing certainly play an important role in the Native American economy, and turquoise has also historically been one of the most versatile resources in Native American history. Its blue color represents the sky and water, elements considered essential for all life. It was ground up into powder form to use as a base for war paint; sometimes consumed, because it was thought to have magical powers; used for various tools; and, worn for decoration.

Today, turquoise is thought of as much more than just a "southwest" stone. It is just as common to see fine turquoise jewelry with diamonds set in gold or platinum as it is to see turquoise set in sterling silver. At times, the worldwide demand has been so high that a lot of simulated turquoise has found its way into jewelry. I also suspect that with the strong economy of the new industrialized country of China and the recent decision to discontinue pegging Chinese currency to the United States dollar, you may see a surge in the cost of turquoise in the immediate future.

Anniversary stone—11th anniversary

Birthstone—December

Composition—Hydrated copper aluminum phosphate

Crystal Structure—Triclinic

Hardness—5 to 6   Toughness—Fair to good

Refractive Index—1.61 to 1.65   Specific Gravity—2.80

Stability: Heat—High heat may damage or discolor stone
               Light—Stable
               Chemicals—Avoid acids, chemicals and cosmetics

Treatments: Impregnation with wax or plastic
               Dying
               Backing with epoxy resin

What causes the blue color: Copper (the green comes from traces of iron)

Sources: China   Iran
               United States

Care and Cleaning: Steam Cleaning—Never
                   Ultrasonic—Never
                   Warm, soapy water—Safe

Even though blue zircon is one of the U.S. birthstones for the month of December, many people seem to avoid this gemstone like the plague. Indeed, if ever there was a gemstone suffering from a public relations problem, it is zircon. Even though it is a natural gemstone, the first words that probably pop into your head are "fake" or "cheap imitation." The reason for this is pretty simple.

Many years ago, colorless zircon was used as a substitute for the much more expensive diamond. Then, as diamond simulate manufacturing became more sophisticated, the albatross of a name seemed to stick. The most common name for a diamond simulate today is "cubic zirconia," and I think most people today think that cubic zirconia and zircon are one and the same.

Zircon
Zircon

The most striking characteristic of the zircon is the high refractive index. What does this mean? The refractive index is a measurement of the fire, or sparkle, that emanates from a gemstone. A diamond (the poster child for sparkle) has a refractive index of 2.42. Very few colored gemstones have anything close to a 2.00 on this R.I. scale. The zircon has a refractive index between 1.93 and 1.98. If you put a blue topaz (which most people use as a December birthstone today) and a zircon side by side, you would see a visible difference in their sparkle. This is the reason I would not accept anything but the blue zircon as my birthstone if I were born in December.

I would classify zircon as a special care stone. Most zircon starts out brown or colorless and is commonly heat treated to bring out its beautiful colors. This heat treatment can make some zircon a little more brittle. If you choose to have zircon set in a ring, I would choose the channel or bezel setting to better protect the stone. Also, never put your zircon in an ultrasonic jewelry cleaner or steam cleaner.

Birthstone—December (blue)

Composition—Zirconium silicate

Crystal Structure—Tetragonal

Hardness—7.5   Toughness—Fair to good

Refractive Index—1.93 to 1.98   Specific Gravity—4.69

Stability: Heat—High heat can cause color change
              Light—Stable, although some treated stones may return to their
                  original color
              Chemicals—Stable

Treatments: Heat

Sources: Australia   Brazil
             Cambodia   China
             France   Myanmar (Burma)
             Nigeria   Sri Lanka
             Tanzania   Thailand
             Vietnam

Care and Cleaning: Steam Cleaning—Never
                   Ultrasonic—Never
                   Warm, soapy water—Safe

COMMON GEMSTONE SHAPES

Round Brilliant   Marquise   Oval   Pear

Heart Brilliant   Princess   Radiant   Octagon

Cushion   Square Brilliant   Triangle   Trillion

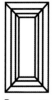 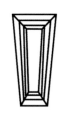

Baguette   Tapered Baguette

DURABILITY SCALES

THE MOHS SCALE OF HARDNESS

1. Talc
2. Gypsum
3. Calcite
4. Fluorite
5. Apatite
6. Orthoclase
7. Quartz
8. Topaz
9. Corundum
10. Diamond

TOUGHNESS GROUPINGS

1. Exceptional- Jadeite, Nephrite

2. Excellent- Alexandrite, Cat's Eye Chrysoberyl, Ruby, Sapphire

3. Good- Agate, Almandite, Amethyst, Bloodstone, Carnelian, Chalcedony, Chrysoberyl, Citrine, Coral, Emerald, Onyx, Pearl, Peridot, Prasiolite, Pyrope, Rhodolite, Rose Quartz, Smoky Quartz, Spessartite, Spinel, Tiger's Eye Quartz, Tsavorite, Turquoise, Zircon

4. Fair- Almandite, Coral, Emerald, Hematite, Lapis Lazuli, Opal, Pearl, Peridot, Pyrope, Rhodolite, Shell, Spessartite, Tanzanite, Tortoise Shell, Tourmaline, Tsavorite, Zircon

5. Poor- Amber, Emerald, Malachite, Moonstone, Opal, Pearl, Tanzanite, Topaz, Turquoise, Zircon

Notice that several gems appear under multiple ratings due to variance in their toughness ratings. For example, Emerald and Pearl can range from good to poor.

ANNIVERSARY STONE LIST

Like birthstone lists, there are a multitude of anniversary lists, and it would be much too cumbersome to list them all. This represents the most traditional of all jewelry anniversary lists.

1. Gold Jewelry
2. Garnet
3. Pearl
4. Topaz
5. Sapphire
6. Amethyst
7. Onyx
8. Green Tourmaline
9. Lapis Lazuli
10 Diamond
11. Turquoise
12. Jade
13. Citrine
14. Opal
15. Ruby
16. Peridot
17. Watches
18. Cat's Eye Chrysoberyl
19. Aquamarine
20. Emerald
21. Iolite
22. Spinel
23. Imperial Topaz
24. Tanzanite
25. Sterling Silver Jubilee
30. Pearl
35. Emerald
40. Ruby
45. Sapphire
50. Gold Jubilee
55. Alexandrite
60. Diamond Jubilee
70. Sapphire Jubilee
80. Ruby Jubilee

STONES OF THE ZODIAC

The tradition of the birthstone was probably derived from the ancient belief that gemstones were of celestial origin. It was believed that these stones possessed magical powers and would bring good fortune to those who wore the stone associated with their zodiac sign.

Like the birthstone lists, there are many lists associated with the signs of the zodiac. Some of them include so many stones that it borders on the ridiculous. It kind of reminds me of television talk shows of today. Who doesn't have one? With that in mind, I have decided to include only the traditional list of the stones of the zodiac.

Aquarius (January 21—February 21)	**Garnet**
Pisces (February 22—March 21)	**Amethyst**
Aries (March 22—April 20)	**Bloodstone**
Taurus (April 21—May 21)	**Sapphire**
Gemini (May 22—June 21)	**Agate**
Cancer (June 22—July 22)	**Emerald**
Leo (July 23—August 22)	**Onyx**
Virgo (August 23—September 22)	**Carnelian**
Libra (September 23—October 23)	**Peridot**
Scorpio (October 24—November 21)	**Aquamarine**
Sagittarius (November 22—December 21)	**Topaz**
Capricorn (December 22—January 20)	**Ruby**

BIRTHSTONE LISTS

There are probably as many published birthstone lists as there are months of the year, if not more. The modern birthstone list compiled by the U.S. Jeweler's Association (USJA) in 1912 is widely accepted as the official birthstone list. However, I have included lists from the American Gemstone Trade Association (AGTA), the International Colored Gemstone Association (ICGA), and the American Gemological Society (AGS).

U.S. Jeweler's Association

January—Garnet
February—Amethyst
March—Aquamarine, Bloodstone
April—Diamond
May—Emerald
June—Pearl, Moonstone, Alexandrite
July—Ruby, Carnelian
August—Peridot
September—Sapphire
October—Opal, Tourmaline
November—Topaz, Citrine
December—Blue Zircon, Turquoise, Lapis Lazuli

American Gem Trade Association

January—Garnet
February—Amethyst
March—Aquamarine, Bloodstone
April—Diamond
May—Emerald
June—Pearl, Moonstone, Alexandrite
July—Ruby
August—Peridot
September—Sapphire
October—Opal, Tourmaline
November—Topaz, Citrine
December—Blue Zircon, Turquoise, Tanzanite

International Colored Gemstone Association

January—Garnet
February—Amethyst
March—Aquamarine, Bloodstone
April—Diamond
May—Emerald
June—Pearl, Moonstone, Alexandrite
July—Ruby
August—Peridot
September—Sapphire
October—Opal, Tourmaline
November—Topaz, Citrine
December—Turquoise, Blue Topaz, Tanzanite

American Gem Society

January—Garnet
February—Amethyst
March—Aquamarine
April—Diamond
May—Emerald
June—Pearl
July—Ruby
August—Peridot
September—Sapphire
October—Opal, Tourmaline
November—Topaz, Citrine
December—Turquoise, Tanzanite

STONES OF THE BREASTPLATE OF AARON

Some of you may have heard about the Bible making reference to twelve gemstones which were set in the breastplate of the High Priest Aaron. These stones were representative of the twelve tribes of Israel. If you read the book of Exodus 39:10—14, it discusses the gemstones used for the breastplate. However, you may not recognize certain gemstone names, and those you do recognize may not necessarily translate to those same names today. As you might expect, there are many different opinions among scholars, theologians, and historians as to what the actual stones were. For example, jacinth has been translated as agate, amber or brown sapphire depending on who you consult. I have taken the liberty of listing these competing points of view where applicable. This list is based largely on interpretations of the original Hebrew as well as what gems were available in the region at that time.

Biblical Name	Modern Name	Hebrew Name/Tribe
I. Sardius	Carnelian	Odem/Reuben
II. Topaz	Peridot or Serpentine	Pitdah/Simeon
III. Chalcedony	Emerald	Bareketh/Levi
IV. Emerald	Garnet	Nophek/Judah
V. Sapphire	Lapis Lazuli	Sappir/Issachar
VI. Sardonyx	Onyx or Diamond	Yahalom/Zebulun
VII. Jacinth	Agate, Amber or Sapphire	Lesham/Joseph
VIII. Agate	Agate (banded)	Shebo/Benjamin
IX. Amethyst	Amethyst	Ahlamah/Dan
X. Chrysoprase	Citrine	Tarshish/Naphtali
XI. Onyx	Turquoise or Malachite	Shoham/Gad
XII. Jasper	Jade	Yashpleh/Assher

GLOSSARY OF TERMS

A.G.T.A.: The American Gem Trade Association

Adularescence: The soft, delicate gleam of color that appears to float across a moonstone.

Alloy: When two or more metals are mixed together.

Asterism: Also known as the star effect. This is caused by needle-like inclusions that are lined up in intersecting patterns. When they cross in the center, they create a star effect. Usually found in rubies, sapphires, or quartz. Depending on the crystal structure of the gemstone, they may be a six rayed star or a four rayed star.

Aventurescence: Sheen or glittery reflections off of small, flat inclusions within the gemstone.This can be found in aventurine quartz and sunstone feldspar.

Bleaching: A process to lighten or remove color. Most common in cultured pearls.

Blemish: Irregularities or imperfections on the surface of a gemstone.

Cabochon: A gem with a rounded domed surface and no facets.

Cameo: The art of carving a gem <u>above</u> the surface in a raised relief.

Carat weight: The standard unit of weight for measuring gemstones, equivalent to 200 milligrams.

Cavity Filling: A process to fill and seal voids in a gemstone to improve appearance and add weight. Sometimes used in rubies, sapphires, tourmaline, opals, and emeralds.

Chatoyancy: Also known as the cat's eye effect. This is caused by needle-like inclusions that lie parallel to one another and reflect light in a narrow band. Most commonly found in chrysoberyl or quartz.

Colorless impregnation: Using wax, plastic, or other substances to fill the pores to improve the appearance and stability of porous gemstones. This may be used in turquoise and jadeite.

Color change: Difference in the body color of a gemstone under different types of light. Most common in alexandrite, but also can occur in sapphire, garnet, or spinel.

Cleavage: The natural way a mineral breaks along certain planes based on its internal crystalline structure. The better the cleavage, the easier it is for the stone to break along those lines.

Density: This refers to the specific gravity of a gemstone.

Dichroism: Gems that show two different colors when viewed at different angles.

Dispersion: Also called fire. The separation of white light into spectral colors.

Doublet: An assembled gemstone made up of two layers.

Dyeing: Adding color to a gemstone to deepen, make the color more even, or change the color altogether. Commonly found in cultured pearls, jadeite, lapis lazuli, mother-of pearl, and turquoise.

Enhancement: The process of improving the appearance of a gemstone. This is also referred to as a treatment.

Facet: One of the flat surfaces of a cut gemstone.

Fracture filling: Filling narrow openings to improve the apparent clarity of a gemstone. This process is very common in emeralds.

G.I.A.: The Gemological Institute of America. The world's foremost authority on gemstones.

Hardness: The ability of a gemstone to withstand scratching.

Heat treatment: Exposing a gemstone to heat to improve its appearance. May be used to lighten, darken, or change the color completely.

Inclusion: Internal irregularities or imperfections in a gemstone.

Inorganic gemstone: A gemstone that comes from minerals, or non-living things.

Intaglio: A design carved <u>below</u> the surface of a gemstone. This is the opposite of a cameo.

Iridescence: A rainbow effect created when light is broken up into many colors. This is most commonly seen in pearls and mother of pearl, where it is called orient.

Irradiation: Changing a gemstone's color through electromagnetic radiation or bombardment with subatomic particles. This process is most commonly used in blue topaz.

Karat: The measurement of the purity of gold.

Lapidary: A technician who cuts, shapes, and polishes gemstones for jewelry.

Luster: A soft reflected light or sheen. Usually used in discussing pearls.

Mohs scale: A scale that rates the hardness of gemstones and minerals on a scale from 1 to 10. This was named after Fredrich Mohs, the mineralogist who invented it.

Nacre: The iridescent substance secreted by mollusks around a foreign object that eventually forms a pearl. This material also lines the inside of the shell to become mother-of-pearl.

Natural gemstone: A gemstone mined from the earth.

Opalescence: A milky white-blue type of iridescence.

Organic gemstone: A gemstone that comes from a once living organism (plant or animal).

Orient: The term that describes iridescence in a pearl or mother-of-pearl.

Play of color: Patches of spectral colors as in an opal.

Pleochrosim: Gemstones that show more than two colors when viewed at different angles.

Points: One point equals one-hundredth (1/100) of a carat.

Refractive index: A measurement of the speed and angle at which light passes through a gemstone.

Rough: Gemstones in their natural state that have not been cut or polished.

Schiller: The play of color in a labradorite.

Simulated gemstone: A piece of jewelry designed to have the appearance of another gemstone. This can be a gemstone (white sapphires used to look like diamonds) or even glass, plastic or other non-gem materials.

Smoke treatment: A surface treatment used to darken an opal and improve its play of color. Low grade opal is wrapped in paper and roasted over fire. The sooty particles penetrate the porous opal and darken the background color.

Specific Gravity: The ratio of an object's weight to an equal volume of water. This is also referred to as density.

Stability: The ability of a gemstone to withstand exposure to light, heat, and chemicals.

Sugar treatment: A surface treatment used to darken an opal and improve its play of color. Low grade opal is heated in a fruit juice solution saturated with sugar. After it cools and dries, it is immersed in sulfuric acid. The sugar converts to carbon and darkens the opal's color.

Surface diffusion: A controversial treatment that uses a combination of chemicals and high temperatures to create a shallow layer of color. The gem material is heated to near its melting point, which allows the chemicals to penetrate the surface and become part of the crystal structure of the gemstone.

Surface modifiers: Treatments used to deepen color or give the appearance of colors on gems. These treatments, which include backing, coating, and painting, have been around for thousands of years. These are the most superficial of all treatments and have no place in the legitimate gemstone world. However, they are still accepted in costume jewelry.

Synthetic gemstone: A laboratory created gemstone that has all of the same physical, chemical, and optical properties of a natural gemstone.

Toughness: The ability of a gemstone to withstand breaking, cracking or chipping.

Treatment: A process applied to a gemstone to improve its appearance or durability.

Triplet: An assembled gemstone with three different layers of material.

Vermeil: A gold coating over silver that must be a minimum of 2.5 microns or 100 millionths of an inch.

Yield: The percentage of finished gemstones that are realized out of a parcel of rough.

RECOMMENDED READING

1. **Joel E. Arem**   *Gems and Jewelry*

2. **Curzio Cipriani & Alessandro Borelli**   *Simon & Schuster's Guide to Gems and Precious Stones*

3. **Daniel J. Dennis Jr.**   *Gems: A Lively Guide for the Casual Collector*

4. **David Federman**   *Modern Jeweler's Consumer Guide to Colored Gemstones*

5. **David Federman**   *Modern Jeweler's Gem Profile/2 The Second 60*

6. **Cally Hall**   *Eyewitness Handbooks Gemstones*

7. **George Frederick Kunz**   *The Curious Lore of Precious Stones*

TOP TEN COLORED GEMSTONES IN SALES

2001

1. Blue Sapphire
2. Pearl
3. Tanzanite
4. Ruby
5. Emerald
6. Amethyst
7. Green Tourmaline
8. Rhodolite Garnet
9. Fancy Sapphire &
   Pink Tourmaline (tie)
10. Blue Topaz

2002

1. Blue Sapphire
2. Ruby
3. Emerald
4. Tanzanite
5. Amethyst
6. Rhodolite Garnet
7. Pearl
8. Opal
9. Peridot
10. Blue Topaz

2003

1. Blue Sapphire
2. Ruby
3. Tanzanite
4. Emerald
5. Amethyst
6. Blue Topaz
7. Tsavorite Garnet
8. Aquamarine
9. Opal
10. Green Tourmaline

2004

1. Blue Sapphire
2. Fancy Sapphire
3. Ruby
4. Tanzanite
5. Emerald
6. Pink Tourmaline
7. Amethyst
8. Blue Topaz
9. Peridot
10. Pearl

Source: Annual survey of U.S. retailers colored stone sales
Colored Stone Magazine January/February 2004 and January/
February 2005

206

TOP TEN COLORED GEMSTONES IN SALES

<u>2005</u>

1. Blue Sapphire
2. Ruby
3. Blue Topaz
4. Fancy Sapphire
5. Amethyst
6. Peridot
7. Tanzanite
8. Emerald
9. Aquamarine
   Citrine
   Opal
10. Rhodolite Garnet

Source: Annual survey of U.S. retailers colored stone sales
Colored Stone Magazine January/February 2006